BECOMING
AN ECOLOGIST

BECOMING AN ECOLOGIST

CAREER PATHWAYS IN SCIENCE

JOHN A. WIENS

Columbia University Press
New York

Columbia University Press
Publishers Since 1893
New York   Chichester, West Sussex

Library of Congress Cataloging-in-Publication Data

Names: Wiens, John A., author.
Title: Becoming an ecologist : career pathways in science / John A. Wiens.
Description: New York : Columbia University Press, [2024] | Includes
    bibliographical references and index.
Identifiers: LCCN 2024033247 (print) | LCCN 2024033248 (ebook) |
    ISBN 9780231218047 (hardback) | ISBN 9780231218054 (trade paperback) |
    ISBN 9780231562041 (ebook)
Subjects: LCSH: Ecology—Vocational guidance. | Ecologists.
Classification: LCC QH49 .W54 2024 (print) | LCC QH49 (ebook) |
    DDC 577.023—dc23/eng/20240814

Cover design: Henry Sene Yee
Cover image: Shutterstock

To those who have shown me the pathways and those who have journeyed with me—teachers, colleagues, students, friends, and family

and

To the nature that we cherish and try to understand

CONTENTS

ACKNOWLEDGMENTS

The experiences I have drawn upon to illustrate pathways to becoming an ecologist span an entire career—a lifetime, really. What I have learned, how I have thought, and what I have done have been influenced by people far too numerous to mention. Many, many students, colleagues, fellow scientists, and friends have shared with me their love of nature and joy of discovery, and my debt to them is great indeed.

My interests and experiences have been diverse. Bringing them together in this book would not have been possible without encouragement and support from Miranda Martin, Brian Smith, and others at Columbia University Press. Lis Pearson's copyedits caught most of my stylistic transgressions, and she tried (with mixed success) to educate me on the proper use of commas.

And most of all, Bea Van Horne has journeyed on these pathways with me for years, tempered my most outrageous speculations, and helped me see many things I otherwise would have missed.

My heartfelt thanks to all.

BECOMING AN ECOLOGIST

INTRODUCTION
Charting Pathways

I'm an ecologist. That's been my professional identity for several decades. Over the years, I've often thought about what I've done and why I did it. What led me to become an ecologist? What forces pointed me in that direction, and what factors determined the pathways I would follow? My answers to these questions reveal a lot about the practice of ecology as a science. They also provide lessons and guidance, I think, for someone contemplating or pursuing a career in ecology.

This, then, is a book about charting pathways of a career in science. It's about how and why one becomes an ecologist, and what one does then.

The steps to becoming a successful ecologist are much the same as those in any area of science. You go to college, declare an undergraduate major, and do well in the required courses. Then may come graduate school and the training and research to obtain a master's or doctoral degree. You may continue with a postdoctoral appointment, but at some point you must get a job that enables you to continue doing ecology. Whether the job is in academia, a government agency, a conservation organization, or something else, you will need to continue with research, obtain grants and funding to support the research, and present the results at scientific meetings and in publications if you want to join the scientific community of ecology. This is a well-established progression. Several books provide detailed guidance for navigating from one step to another—for example, how one becomes a conservationist, neuroscientist, or paleontologist.[1]

This is not such a book. To be sure, my narrative illustrates these steps to becoming an ecologist, but this progression only describes *how* one becomes an ecologist. My focus instead is on the pathways one might follow along the way and *why* certain pathways are (or are not) followed at particular times. For many ecologists, career pathways do not follow a straight-line route planned well in advance, smoothly unfolding before them as they go. Nor are there always well-defined intersections where one can choose to go one way or another. Instead, there are multiple interweaving, diverging, and converging pathways, rather like a braided stream making its way through a floodplain.

Understanding why pathways wander so says much about how and why someone becomes an ecologist. What forces, factors, and events determine the pathways that you follow? Why do you follow some pathways but not others? Which pathways you follow determines what kind of ecologist you'll be (or if you'll become something else). Charting pathways can also reveal a good deal about how ecology has developed as a science. Even more broadly, it depicts how scientists actually work. Why do they ask the questions they do, and how do they go about answering them? Why do they expect certain answers to their questions, and what do they do when the answers are not what they expect?

So this is a book about pathways to becoming a scientist—an ecologist.[2] Before we get too far, however, I should explain what I mean by "ecology" and "ecologist." As the discipline of ecology has grown and become more relevant to more people, "ecology" has taken on multiple guises and meanings.[3] Not only are there different areas of specialization within ecology depending on what is studied (e.g., aquatic ecology, systems ecology, landscape ecology, microbial ecology, small-mammal ecology), but ecology is also part of other disciplines such as forestry, wildlife management, or limnology. Anthropology, economics, and other social sciences often include elements of human ecology.[4] Ecology underlies activist movements concerned with water or air pollution, urban sprawl, deforestation, or the erosion of biological diversity—"saving the environment."[5] "Ecology" has become a catch-all term for anything having to do with the environment. Accordingly, there are many kinds of ecologists and many ways to become one.

My use of "ecology" and "ecologist" in this book is more traditional and restricted. As I use it, "ecology" is a biological science that seeks to understand how living organisms relate to their physical environment and to one another, and the consequences of those interactions. Ecology strives to unravel the complexities and interrelationships of nature. An "ecologist," therefore, is one who takes a scientific approach to such studies.

~~~~

Pathways of a career in ecology, or any scientific discipline, have many twists and turns that are influenced by an interplay among several forces and factors. These are the recurrent themes of *Becoming an Ecologist*, so I'll highlight them here. It starts with *interests*. Individual interests and curiosity set the stage for embarking on pathways. Scientists are often drawn to study things that excite them—they have a passion for the predictability of physics, the grandeur of astronomy, or the mechanistic tidiness of molecular biology. Most ecologists have a passion for nature. These interests are often unfathomably deep. They may be kindled in childhood or spring to life later, perhaps sparked by a chance encounter on a family vacation, a book one has read, or an inspiring teacher. Although the pathways followed by a scientist may change over time as interests come and go, the core of one's interests may persist for a lifetime. The interests and passions that emerged in my own childhood—being out in nature, a love for open spaces, an attraction to birds, a fascination with how places and environments varied, and (eventually) writing about all of this—remained, no matter what pathway I was on.

Following a pathway also requires *motivation*. Motivation is what leads you to explore a topic and ask questions, guided by your interests. Motivation keeps you plugging away at a question when the answer eludes your grasp. Motivation prompts you to start down a pathway and determines how long you persist or whether you switch to another pathway. My own progression was driven by what motivates many scientists—an abiding quest to find things out, to solve nature's puzzles.

Interests and motivation determine whether a pathway is likely to be followed. But *opportunity* determines whether it *can* be followed, whether the portal to a pathway is open to you. Scientific training and experience prepare you to take advantage of opportunities to embark on a career pathway that may have been planned well in advance. Other opportunities, however, may appear almost by chance and without warning, unplanned consequences of events beyond your control. And you can spend a lifetime waiting for an opportunity that never appears. When an opportunity that aligns with your interests and sparks your motivation does appear, it can trigger a shift to a different pathway.[6]

Interests, motivation, and opportunity are the driving forces that bring one to the portal of a pathway. Several additional factors influence the decision of whether to embark on a certain pathway. The *people* one encounters, for example, can play a pivotal role in determining how a scientist develops, and which pathways are followed. Parents, relatives, teachers, and friends may arouse an interest that becomes a passion. Graduate advisors direct the training of scientists and help reveal pathways you didn't even know were there. Colleagues can pull (or push) one onto new pathways. People can inspire a scientist to follow—and persist in following—a chosen pathway or, sometimes, to abandon a pathway that is likely to lead nowhere. Influential people can create opportunities, opening portals to new pathways.

Pathways are also influenced by *societal relevance*. What scientific research does society deem acceptable or important? Should science address the needs of a community of users such as foresters, farmers, or fishers? What questions merit societal support? Are the findings important only to other scientists or do they have broader societal relevance? Pathways that are guided by societal relevance may lead one to ask different questions and pursue different sources of funding.

Of course, scientists are people too, and the *practicalities of life* influence their choice of pathways. The demands of family and job, for example, may create expectations or impose restrictions that open some pathways but foreclose others. Where you grow up and where you live and work may affect how your interests develop, the people you encounter, and the societal culture in which you are embedded. Places have a powerful effect on which pathways are available and which are followed. Changes in location, whether by design or serendipity, can create opportunities to embark on new pathways or make following a current pathway no longer feasible.

Finally, the *culture of a science* can exert a powerful pressure to follow some pathways but not others. In *The Structure of Scientific Revolutions*, Thomas Kuhn offered an argument for why this might be so.[7] Kuhn called the overarching framework of methods, observations, hypotheses, and theories in a science a *paradigm*. He suggested that a paradigm directs and circumscribes investigations in a discipline. This sets the stage for "normal science," in which adherents to the paradigm form a tight community of like-minded scientists.[8] The paradigm determines what one's scientific peers find interesting or important, what questions are worth pursuing or should be abandoned, and what research is likely to be funded and published. The paradigm becomes a central part of the scientific culture.

Inevitably, however, some observations may emerge that don't fit the paradigm. Such anomalous observations may initially be ignored or explained away, but eventually enough exceptions may accumulate to challenge the explanatory power of the paradigm. Confidence in the paradigm is shaken and challenges to the paradigm become more frequent. This may open the door to the emergence of a new framework that explains the anomalies. Scientists may then rally around a new paradigm, launching a new phase of normal science.

Kuhn's view of the scientific process has been criticized,[9] and it may be a bit of a stretch to apply it to ecology.[10] However, I believe Kuhn's model of the scientific process is a useful way of thinking about the culture of ecology as a science. Some concepts in ecology have become so widely accepted that they function as paradigms, guiding how many scientists ask questions and conduct research. As in Kuhn's phase of normal science, studies that align with prevailing thinking may be favored over those that don't. Anomalous observations may be ignored, but they may also lead you to ask different questions or change research directions. Kuhn's model provides some insight into how the culture of ecology can act as a force determining how (or whether) you progress along a pathway.

My aim in *Becoming an Ecologist* is to show how and why these driving forces—interests, motivation, and opportunity—determine whether, and how, a scientist comes to a pathway and how the factors—influential people, societal relevance, life's practicalities, and the culture of a science—affect the decision to follow it. In the process, I want to shed some light on how ecology as a science is actually done. I use my own experiences as an ecologist working in several scientific disciplines—ornithology, community ecology, landscape ecology, and conservation—to illustrate how these forces and factors influenced the weaving trajectory of a particular career pathway—my own. Rather than simply recounting some of what I've done as a scientist, however, I dig deeper to reflect on *why* I did what I've done. I do this because my career provides clear examples of the influences of these forces and factors (some of which I recognized only in retrospect), and because it is (obviously) the career pathway I know best.

My narrative is punctuated by a series of episodes. The episodes show how and why I followed certain pathways and not others and asked certain questions and not others. They reveal how pathways are shaped and directed by a mix of internal and external forces, with a large dose of serendipity and chance. My narrative illustrates how ecologists go about studying nature and how nature works. It offers a perspective on some past debates and current issues in ecology as a science and its applications to resource management and conservation. Because it has spanned decades, my career trajectory also provides insights into how the currents of history unfolded in several disciplines.

I was lucky. My early interests in birds and nature were nurtured and allowed to develop freely, so when opportunities came along, I was primed to take them. And I felt that I could take advantage of opportunities to shift pathways without derailing my career. As a white male I also had opportunities and access to pathways that, at the time, were not available to everyone. When I began my first faculty position in the late 1960s, for example, several senior faculty members counseled me not to accept female graduate students. They can't take the rigorous demands of ecological fieldwork, I was told, and they'll probably just get pregnant so all your efforts in training them will be for naught. Applicants for graduate studies were required to include a photograph with their application. Years later, female applicants for faculty positions in some male-dominated wildlife and forestry departments still faced greater scrutiny and endured more aggressive grilling than their male counterparts, and those who were hired often faced discrepancies in job expectations and salaries. The challenges faced by Blacks or Hispanics could be even greater, and LGBTQ+ individuals learned to keep their lips tightly sealed. Societal changes and shifting attitudes have made opportunities and pathways in ecology available to a more diverse array of people, but there are decades of catching up still to be done.

The particulars of my journey reflect the times in which I made it, from the middle of the last century to the first decades of the present century. Ecology changed over this time. Conservation biology and landscape ecology emerged as new areas of study that applied ecology to issues of public concern. The internet now provides a ready source of information instantly available on nearly everything. An exploding array of tools and techniques generates masses of data, and artificial intelligence can sort through the masses looking for interesting patterns.[11] Ecology now has a well-established body of theory ("principles," or even paradigms) that helps to determine what is interesting, to whom. Ecology has grown to become part of other scientific and sociological disciplines. Today's ecology is quite different from what it was when I started out, and much of my journey reflects my attempts to keep up with the changes. "Becoming an ecologist" is an ongoing process; it's not finished when you get a PhD.

My narrative in *Becoming an Ecologist* describes what ecologists (this one, at least) do, how they think, and why they follow the pathways that they do. But is any of this relevant to someone contemplating a career in ecology now? Is it all ancient history—interesting perhaps or maybe even inspiring—but not a road map that can or should be followed?

The details of my journey to become an ecologist are mine alone. The pathways followed by my contemporaries were all different and individualistic, and those followed by people entering the field today would be different still. Yet the driving forces and factors that determine the trajectory of career pathways to become an ecologist remain. By focusing on the forces and factors that have guided my pathways and how and why I responded to opportunities to shift to different pathways, I want to show how following one's interests and passion for the natural world can be the foundation for a career in ecology.

1

IN THE BEGINNING
Emerging of Interests

Scientists are curious. They want to find things out. They ask many questions, for many reasons. Sometimes the questions arise unbidden from something they've seen or heard. Sometimes they are prompted by the predictions of a broader theory or the premises of a paradigm. Usually, however, scientists ask questions about things that interest them.

Interests spark the internal fire that keeps a scientist going, often pursuing a question until an answer emerges or the question mutates into something different, opening up new possibilities. Interests are what wake a scientist up in the middle of the night to write down thoughts that may hold the answer to the question (or may, in the morning light, be gibberish). Interests are what underlie "the pleasure of finding things out."[1]

Once an interest has been kindled, following that interest down a pathway may be reinforced by other forces. The culture of a scientific discipline may dictate what other scientists will find interesting. Broader societal forces may encourage, discourage, or even forbid certain interests. And chance events may provide opportunities that align with interests or restrict options.

Interests are often established in childhood. Most children have an interest in something. For some, it may be a single-minded, burning passion. Others may have more eclectic interests, flitting from one thing to another like butterflies to flowers, depending on what strikes their fancy at the moment. In either case, the interests developed in childhood often shape the pathways followed later in life. There are doorways or portals to many possible pathways; childhood interests often determine which ones open and which remain closed or are not even seen.

Where do a child's interests come from? Why do some children have an insatiable interest in beetles, or rocks, or oddly shaped sticks? In some cases, it's simply because that's what interested their parents, siblings, or friends. The environment is also important. Children growing up in a city are exposed to stimuli and opportunities that those living in the countryside lack (and vice versa), piquing some interests but not others. And economic circumstances may deny a child the opportunity to develop interests that otherwise might grow and flower. Childhood poverty has many effects, but among the most devastating may be the restriction or suppression of interests.

One's interests are not immutably fixed, of course, nor are they always established in childhood. Experiences at a summer camp, an inspirational college professor, a personal crisis, a chance encounter—these and other factors may spark an interest that sets someone on a pathway to becoming a scientist. Yet for many scientists, perhaps especially ecologists, early experiences and childhood interests play an outsized role in pointing the direction to a later career pathway. Charles Darwin's interest in nature was kindled at an early age, when he would take long walks into the English countryside to watch birds and collect insects, rocks, or whatever else he could find. Many of today's ecologists were birdwatchers or butterfly collectors when they were young. In his autobiography, *Naturalist*, E. O. Wilson recounts how his childhood interest in ants led to a career that spawned many advances in population biology, evolution, and ecology.[2] The interests of these people in nature took them outside, where they became budding naturalists. Wilson makes much of never abandoning the fascination with the natural world that struck him first as a child.

As technology has become more enticing and there are fewer wild places to explore, however, it has become more difficult for children to experience nature as Wilson (much less Darwin) did. In his book, *Last Child in the Woods*, Richard Louv drew attention to the effects that a diminished experience with nature can have on the emotional and physical health of children (and adults)—what he termed "nature-deficit disorder."[3] This may especially affect the opportunity for children to follow an interest in nature onto a pathway to ecology. A few years ago the Canadian ecologist Charles Krebs and I wrote an essay, *The Last Ecologist in the Woods?*, urging ecologists to follow the admonition of the nineteenth-century naturalist Louis Agassiz to "study nature, not books."[4] The diminishment of opportunities to experience unspoiled nature is real, but this doesn't mean that curiosity about the natural world cannot be nourished. There are beetles to be found under logs and (as Wilson found) ants to be discovered in vacant urban lots. And birds are everywhere.

Being an ecologist, however, is hard. The Holy Grail of many sciences is to derive general laws that apply broadly, as in physics or mathematics. "General laws," or even satisfying generalizations, are elusive in ecology. The nature that ecologists study is complex and varies at multiple scales. Every time, place, and assemblage of organisms seems to play by different rules. Moreover, for centuries people have abused and eroded the webs of interrelationships that are part of the beauty of nature. As Aldo Leopold (who had much to say about ecology and nature) observed, "one of the penalties of an ecological education is that one lives alone in a world of wounds."[5] The challenges are daunting. Of course, you can have an interest in nature without being an ecologist (many people do), but I maintain that it's difficult to be an ecologist without having a deep interest in nature. If you want to be an ecologist, then, you should begin by identifying what *really* interests you. Then consider which career pathways might best align with your interests.

So where did my own early interests come from, and how did they affect my later choice of career pathways? What led me to become an ecologist? The stage was set by an intersection of two early interests: exploring nature, and birds. Until I was six years old we lived in Idaho, where my father taught German at the University of Idaho, in Moscow. Some of my most vivid childhood memories are of the times we spent at a cabin deep in the dark and forbidding forest (or so it seemed to me) on Moscow Mountain. My father and I explored, catching frogs in the streams or finding salamanders or millipedes under overturned logs. In the summers we visited my grandfather in Bethany, West Virginia, where we'd discover hidden places in the hills and ravines. I learned that bats lived in abandoned coal mines and that creeks were full of "crawdads" (crayfish). It was all a great adventure, and I loved it.

This was in the early 1940s. As World War II came to a close, the G.I. Bill provided benefits for returning veterans to attend college. Enrollments exploded and universities hired new faculty to keep up. In 1946, my father joined the faculty at the University of Oklahoma and we moved to Norman. That's when I was first exposed to discrimination and social inequities. When we lived in Idaho during the war I had been told that the Japanese were the enemy, and I heard my parents mention the internment camps in California, Oregon, and Idaho. But Moscow was an academic-agricultural community with few minorities. I was a child, blissfully unaware of such things. Oklahoma was different. Racial discrimination was rampant. The Tulsa Race Massacre had happened only a quarter-century before (although it was never mentioned in my schools and I learned about it only much later). There were separate (dingy and dirty) water fountains and washrooms for "Coloreds Only" at the train station in Oklahoma City. And gender roles had become ossified after the war. Women were expected to stay in the home while their husbands had jobs and pursued careers. It was a world according to *Dick and Jane* readers or television shows like *The Adventures of Ozzie and Harriet*. My parents decried the discrimination they saw and championed equal opportunities for women, but only around the dinner table. They were not activists. This was a time of McCarthyism and my father, an immigrant from Russia and a professor of German, thought it best to keep a low profile.

At the time, I had little direct experience with these issues. Instead, my interest in birds was beginning to emerge. As often happens with children, this interest rapidly became a passion. By the time I was nine, I had saved up and bought some heavy and clunky binoculars. I made a bird feeder and installed it outside my bedroom window. I learned how to identify birds and kept records of what I saw. I joined the Cleveland County Bird Club and attended their monthly meetings. I went on their birdwatching field trips. By the time I was eleven, I was riding my bike into the countryside, searching for bird nests, following their fates, and keeping careful notes. I had become a boy birdwatcher.

Why birds? Why not stargazing or building model airplanes or something else? I really have no idea. The interest just seemed to be there, waiting to be released and take flight. Perhaps it was because birds were colorful, sang appealing songs, and were conspicuous and plentiful. These traits make birds interesting to many people, but not usually a consuming passion. Whatever the reasons, my interest in birds, once it emerged, overwhelmed other interests (at least for a while).

My delight in being outdoors and interest in birds intersected with a strong attraction to open spaces. Norman lay at the transition between the woodlands that extended east of town and grasslands and prairies to the west. When I headed out into the countryside on my bike, I almost always went west. I was more comfortable in the prairies; I could *see* things without all those trees getting in the way. This attraction to open spaces would later shape much of my scientific research.

Emerging childhood interests are molded by one's family, in several ways. First, to pursue something with a passion requires strong motivation, particularly during childhood when so much of the world is new, exciting, interesting, and potentially distracting. Both of my parents were role models of a fierce passion for learning and teaching and the importance of motivation in reaching one's goals. My mother was motivated to play basketball in high school at a time, in the 1920s, when women's sports were just beginning to take off. She later earned a master's degree in English from The Ohio State University, writing a thesis on the works of Edgar Allen Poe. When we moved to Oklahoma, she took a position as an instructor in the English department at the University of Oklahoma to satisfy her passion for teaching (and earn some extra income). Teaching ran deep in her side of the family. Her sister taught primary school in Bethany, her brother became an English professor at Stephens College in Missouri, and her father received a Master of Divinity degree from Yale and for many years was pastor (i.e., a teacher) of a church in Bethany.

My father's background was entirely different. He grew up as part of a large family in a small farming village that was part of the Molotschna Mennonite colony in Ukraine. Despite his farming background, he had a passion for literature and the arts. He endured famine and war following the Russian Revolution and was allowed to immigrate to Canada during a lull in the fighting in 1924. At first he worked on farms but then, after a stint of teaching in a one-room schoolhouse in Manitoba, he was able to pursue his academic interests. This led to a master's degree in his new language (English) and a college scholarship. He completed a PhD program in German at The Ohio State University in 1934 and became a university professor. This wouldn't have happened had he not been highly motivated.

Second, I was (and still am) an only child. Only children have a variety of character traits, which I won't go into here. Lacking siblings, however, meant that I had no one close with whom I could share my interests and enthusiasms. I fear that I became (and perhaps remain) somewhat of a loner. Consequently, when I rode my bike into the countryside or searched for bird nests, I did so alone. My childhood friends were interested in stamp collecting or sports, but not in the outdoors or birds. The seeds of an independent (some would say contrarian) streak were planted.

Third, I did not grow up in a family of scientists. My parents were both university professors in the humanities. No one in my family lineage, going back generations, was even close to being a scientist. Our house was full of books, but all of them were about art, literature, philosophy, history, and language—not science. My friends were the sons and daughters of other academics, none of them scientists. My favorite classes in high school were in English literature. It did not occur to me that my interest in nature and birds could lead to a pathway to becoming a scientist.

My father, however, did love nature. We often took hikes together just to explore. We searched for big trees, which fascinated him, perhaps because large trees had long since been felled in the steppes where he had grown up. He knew little about the plants and animals we saw; he just wanted us to be outside. I think this fostered in me an appreciation for the broad sweep of nature; I would fill in the particulars later.

By the time I turned ten, my interest in nature and birds was well-established. But childhood interests often need a boost at some point. Mine came through a family connection. In the fall of 1952, George Miksch Sutton joined the faculty at Oklahoma. Sutton was an internationally acclaimed ornithologist, bird artist, writer, and naturalist. He was also a close friend of our family. Sutton had stayed with my grandparents while he completed an undergraduate degree at Bethany College.[6] He often took my mother and her siblings on his excursions as he scoured the hills to collect birds for the Carnegie Museum in Pittsburg, where he was an assistant curator. My mother had spoken of these times, so we reconnected with Sutton when he moved to Norman. His tales of expeditions to Mexico to collect and paint birds fascinated me. He showed me where an interest in birds and the outdoors could lead.

My passion for birds played into Sutton's hands. He showed me the bird collections at the university's Stovall Museum and explained how scientists used them to classify birds and document their distributions. Sometimes he took me along on field trips. When Margaret Morse Nice visited him shortly after he came to Norman, he arranged for me to go along with her on a weekend trip to the Wichita Mountains in southern Oklahoma. Margaret Nice was an ornithologist who had studied the distribution of Oklahoma birds in the 1920s and was well-known for her definitive study of the life history of the song sparrow.[7] We camped in the mountains for two days, hiking the trails while she pointed out plants, insects, and (of course) birds. I watched as she sketched a drawing of a fence lizard that I had caught. It didn't occur to me then how unusual it was, at that time, for a woman to be a well-known scientist and ornithologist, but she was an inspiration to me years later when I began directing female graduate students.

Sutton often asked his graduate students whether they would mind having a thirteen-year-old boy tagging along with them as they gathered data for their dissertation research. What could they say? One of his students, Chris Johnson, was studying nesting birds in a grassland site near Norman. He took me along one day, showing me the cleverly made domed nests of grasshopper sparrows and meadowlarks. It must have made quite an impression on me, as I was later to spend nearly a decade conducting research on grassland birds.

When I was sixteen, I took a summer job at a dude ranch in Colorado. Sutton wrote to me about the challenges of striking out on your own (even for just a summer), while also urging me to, "By all means keep up with your bird study."[8] Nevertheless, he was careful not to push me; he wanted my interest in birds to develop on its own. It persisted for a while, under Sutton's tutelage, but then began to wane. When I entered high school, my interests shifted from birds to basketball and girls and from natural sciences to humanities and the arts. Literature and writing grew more important to me. I became editor of our high school newspaper. My best friend was an English literature geek who distained athletics and had little use for science.

With the social demands of high school, I no longer had time for lonely trips into the countryside. Birdwatching wasn't cool. My interest in birds and the outdoors was still there, however, reinforced by occasional outings with a close friend, David Timmons. David shared my interest in the outdoors, but as a Native American—a Cherokee—he had a different, more contemplative way of looking at nature than I did. When David and I went out together, we *absorbed* nature rather than simply observing it. The lessons were later to be useful when I needed to temper routine data collection with patient and open-minded observations.

At this point, I had little understanding of science or appreciation for the scientific process. In fact, my high school (a small, university-operated school) offered only one science course, "General Science." It was taught by a Baptist fundamentalist who didn't believe in evolution. Most of the class was about physics, geology, the Great Flood, and Special Creation. Somehow my latent interests survived; I must have been highly motivated. But I still imagined following the pathway of my parents and relatives toward a career in the humanities, perhaps as an English professor specializing in the works of Hawthorne or some such. My future seemed clearly *not* to be in ornithology or ecology, or in any area of science. Perhaps birdwatching would be a pleasant hobby.

Yet all the while, George Sutton patiently waited.

2

MOLDING OF INTERESTS

My interests in nature and birds and my love for open spaces were established in childhood. They were nurtured as I grew older by several people, particularly my father and George Sutton. But by the time I graduated from high school and began my freshman year at the University of Oklahoma (OU), these interests had been submerged or sidetracked. I had abandoned any notion of becoming an ornithologist for my life's work. That was not a pathway I thought worth following. What caused these interests to reemerge and then be molded into the beginnings of a career in science?

I wasn't really sure what I *did* want to do. I started out in petroleum engineering. After all, this was Oklahoma, where oil was king. There would be many employment opportunities after I graduated. I spent the first term in a required course devoted to teaching one how to use a slide rule (this was how engineers did their calculations then, I guess). That was enough to tell me that my future pathway lay elsewhere. So I switched to the "University College," a home for undecided students who were still searching for a pathway. For two years I bounced around, taking an eclectic mix of courses and declaring an undergraduate major in whatever seemed like a good possibility at the time. I tried journalism because I enjoyed writing. But the professor insisted that the only way to write was in simple, one-sentence paragraphs. I've since realized that journalists may produce some of the most eloquent and moving writing to be found anywhere, but that's not what we were being taught. I tried English, thinking that would mesh more comfortably with my interest in writing, but the classes seemed to be mostly about how other people wrote. Then I tried philosophy, which was great intellectual fun, but I couldn't see it leading anywhere. Russian studies were much in vogue then, but I simply couldn't master the language. I gravitated back to English again.

At some point in all this hopscotching among majors I realized that I would have to take a science course to graduate, whatever my major. So that spring I signed up for Introductory Zoology. This was a turning point. To get to know his students, the professor, Robert Burns, asked each of us to fill out a form indicating (among other things) what magazines we read. Despite my flagging interest in birds, I had maintained my membership in the Wilson Ornithological Society, so I put down *The Wilson Bulletin*. Bob thought a scientific ornithological journal was an odd choice for a college sophomore at OU. He took an interest in me, invited me to join in on field trips with his mammalogy class, and guided me toward majoring in zoology. I was excited and focused at last. I quit bouncing around among majors.

All this while George Sutton had been waiting in the wings for me to make up my mind. When he learned that I had committed to a zoology program of my own free will, he was delighted. He wrote to me where I was working that summer as a stockroom manager at the Colorado dude ranch, to offer me a job. I would be his research assistant. I would help him document the details of the distribution and systematics of the birds of Oklahoma, something that had interested him since his first expeditions to the state in 1932. My responsibilities would involve maintaining files of distributional records and "some field work, some bird skinning, a good deal of curating, and helping me to keep things neat."[1] It was an unexpected opportunity, and I jumped on it.

The work started out in Sutton's office and the university museum. It wasn't long, however, before he involved me more directly in the fieldwork to help fill gaps in the distributional records for Oklahoma birds. This meant collecting and preparing specimens. My metamorphosis from a birdwatcher to scientist began with collecting.

EPISODE: THE COLLECTOR

Ornithological research now relies on photographic records to verify distributional observations and genetic analyses to differentiate variation within a species. It was quite different in 1960. Distributional records had to be supported by a documented physical specimen; sight records or even photographs were susceptible to misidentification and wouldn't do. Systematics, particularly the determination of variation among subspecies, was largely based on measurements of morphological features of museum specimens. Although George Sutton was a consummate naturalist, his fieldwork in Oklahoma revolved around collecting—shooting birds and preserving specimens as museum study skins. So I learned how to shoot (my father, never a hunter, distained guns, perhaps because of his experiences growing up during the Russian Revolution). Sutton taught me how to skin a bird and prepare a stuffed specimen that wasn't so grotesque as to be unidentifiable.

By my senior year in college, I was thoroughly taken by documenting the distribution and systematics of birds. I took Sutton's ornithology classes. I studied subspecific variation among species and measured specimens in the museum collections. I went on field trips with Sutton's students as they collected specimens for their graduate research. Often, I was joined by David Ligon, a classmate who was also a budding ornithologist and a Sutton assistant. The two of us spent most of our spare time that year (including some time we should have been in classes) in the field, searching for and collecting birds to fill in gaps in the distributional records for Oklahoma birds.

Oklahoma is a biogeographically diverse state, straddling the east-west transition from forests to prairies. One corner of the state touches on the swamps, bayous, and pine forests of the Southeast, while the northwestern tip of the Oklahoma Panhandle juts into the high plains and mesas on the eastern flanks of the Rocky Mountains. The northeast part of the state is part of the Ozark Plateau, and the southwestern corner includes gypsum hills that extend into western Texas. Our collecting trips took us to all these quarters.

Collecting songbirds, even for scientific purposes, runs counter to the sensibilities of many people. Even in the 1930s, birdwatchers had been concerned that Sutton would collect rare birds on his expeditions to Oklahoma and take them back to eastern museums.[2] He insisted that we check with landowners to explain what we were doing and why it was important. Sometimes it took a bit of embellishment to convince doubting landowners to allow us on their land. In Oklahoma, football was a religion. Although I bear scant physical resemblance to a football player, I took to mentioning that I played sports at OU (omitting the detail that it was intramural basketball). This often opened some sticky gates.

David and I usually knew what we were looking for when we visited an area and were constantly on the lookout for species that "shouldn't be there." We took several trips to the Black Mesa at the tip of the Oklahoma Panhandle, looking for western bird species that didn't make it into the rest of the state. On one of these trips, we were searching among the cottonwoods on the floodplain of the Cimarron River. I spied a lesser goldfinch in a treetop. Knowing that it was needed in the museum collections, I raised my gun and fired. The bird fell to the ground. Then things got interesting.

Unbeknownst to us, the Oklahoma Ornithological Society had scheduled their spring field trip to the Black Mesa country. At that moment, some twenty birdwatchers were on the other side of the tree, looking at the goldfinch as it fell to the ground. As I picked it up, I saw them, binoculars lowered and mouths agape, staring at me coldly. What could I do? There was no place to hide. I decided to make it an educational moment. I went over to the group and showed them the bird, pointing out distinctive field marks and plumage characteristics. I explained that we were there to collect birds for the OU museum, to further our understanding of the distribution of birds in Oklahoma. I mentioned that we were working with George Sutton. They all knew who he was. Some of them shared the adverse reaction to collecting that Sutton had encountered on his expeditions to the same area thirty years before. We parted amicably enough, but David and I were careful to avoid the group for the rest of our trip.

My experiences as a collector solidified my interest in birds, particularly their distribution and systematics. George Sutton was putting together the definitive work on the birds of Oklahoma, and David Ligon and I were helping find missing pieces of the distributional puzzle.[3] I spent many hours in the OU museum looking at study skins, measuring wing lengths, and recording the subtleties of plumage variation to satisfy myself that the subspecific designations were accurate. I decided that distribution and systematics would be my future focus, probably pursued through a museum. A pathway was taking form.

Looking back, it's easy to be embarrassed, or even a bit shamed, by my enthusiasm for killing birds so they could be preserved as study skins in museum trays. Collecting is no longer the centerpiece of museum work as it was in the 1960s, much less as it had been decades earlier when George Sutton was beginning his career. The value of museum collections has only grown, however, as new approaches to avian systematics and new technologies for analyzing the minerals and DNA in feathers have developed. At the same time, the network of bird enthusiasts contributing information has expanded almost exponentially. Citizen science efforts such as eBird document distributional changes and population trends at local to continental scales.[4]

Although I continued to collect specimens for research projects for several years, it has now been decades since I collected a bird. As my interests shifted, I found it much more satisfying to watch what they were doing. But I've never regretted my time as a collector for Sutton and the OU museum. Searching for elusive species took me to places I otherwise never would have seen. In the process, I learned about the habitat associations and behavior of species, where to look for them, and how to approach them. I didn't just collect them; I *observed* them. To be a collector one must also be a naturalist.

When I finally decided that the pathway to fulfilling my interests in birds went through zoology, I realized that I hadn't been taking the requisite coursework for a degree. I had some catching up to do. So I signed up to spend the following two summers at the University of Oklahoma Biological Station, where I could bolster my transcript and rub elbows with kindred souls.

The OU Biological Station was on the shore of Lake Texoma, a large impoundment of the Red and Washita rivers on the Oklahoma-Texas border. The surrounding woods and fields provided plenty of opportunities to satisfy my yearnings to spend time in the field, and all of the courses I took had a large field component. But I was also encouraged to undertake an independent research project, beyond what was required in class. I wanted to do something with birds, of course. It seemed only natural to do what I'd been doing since childhood—find nests and follow their fates to see what happened. When I wasn't in classes I was roaming the fields, watching birds, finding their nests, and taking detailed notes on nest placement, surrounding habitat, egg-laying, and success (or failure) in fledging offspring. It wasn't as easy as I expected. Summers in southern Oklahoma can be blisteringly hot, and it was muggy near the lake. The birds seemed reluctant to divulge the locations of their nests. And my encounters with snakes were all too frequent.

Time for a short digression. I have always had a difficult time with snakes. I'm not sure I would call it fear as much as an inordinate respect. Among the classes I took at the OU Biological Station was herpetology. It was taught by Charles ("Chuck") Carpenter. Carpenter relished going into swamps to ferret out water moccasins or exploring rocky outcrops to catch rattlesnakes—the bigger, the better. Most of the other students in the class were similarly inclined. I wasn't. I was perfectly content to find a snake and then call one of the others over to catch and handle it. As the end of the term approached, Carpenter pulled me aside. "John," he said, "I've noticed that you've done fine on lizards and frogs, but you haven't caught any snakes yet. Just so you know, if you don't bring me a snake you've caught you won't pass the class." So warned, I worked up my courage and managed to catch a young garter snake—about as docile and unthreatening as a snake can be. I proudly showed it to Carpenter. With a begrudging smile, he admitted that I'd kept my part of the bargain. I passed the course. But I still avoid snakes whenever I can.

EPISODE: COWBIRDS IN OKLAHOMA

Despite the heat and the snakes, I was excited to actually be doing research—finding things out. As I found nests, however, I discovered that some were parasitized by brown-headed cowbirds. Cowbirds are one of several bird species that are brood parasites. Rather than building nests, laying and incubating their own eggs, and rearing their hatchlings as proper parents do, cowbird females surreptitiously lay eggs in nests of a host species. In some cases, the hosts are not easily duped: the alien egg may be removed, the eggs may be covered with new nest material on which the host lays another clutch, or the host may abandon the parasitized nest, perhaps to try again elsewhere. But many host species are defenseless. The unsuspecting host may accept the cowbird egg, hatch it, and rear the chick, probably wondering what went wrong as the cowbird chick grows faster and becomes larger than the host chicks. The cowbird chick can monopolize food and even evict the host chicks, so it has the nest all to itself. Where cowbirds are abundant and hosts lack defenses, host nesting success can suffer. Cowbird parasitism is a major contributor to the imperiled status of several critically endangered bird species, such as Kirtland's warblers and black-capped vireos. Conservation efforts for these species have centered on removing cowbirds from nesting areas by trapping them and on maintaining habitat by prescribed burning. These measures have been successful, and both species are no longer listed as endangered. Were the cowbird control and habitat management efforts not to continue, however, the population gains would be reversed. Both the warbler and the vireo require ongoing management—they are what several of us have called "conservation reliant" species.[5]

Cowbirds were common in the agricultural setting of southern Oklahoma. It turned out that many of the nests I found contained a cowbird egg. Although I hadn't started out with well-defined questions (much less a guiding theory or hypothesis), I now began to structure my observations around some questions: What was the incidence of parasitism and how did it differ among host species? Did cowbird parasitism affect the breeding success of hosts?

I spent two summers finding nests and following their fates. Overall, I found over fifty nests of eight species, nearly two-thirds of them parasitized. Some species were more attractive as hosts than others. All of the cardinal and orchard oriole nests I found were parasitized, while only a third of the dickcissel nests contained a cowbird egg and none of the red-winged blackbirds nesting in a nearby wetland was parasitized. Parasitized nests were equally successful in fledging cowbird and host young (24 percent and 19 percent, respectively), while fledging success was greater in nonparasitized nests (36 percent) of the same host species. So cowbird parasitism clearly had an impact on at least some hosts.[6]

As I assembled my observations at the end of the study, another pattern emerged. Even though the timing of egg-laying by host species overall declined steadily from a peak in early June until late

July, egg-laying by cowbirds in host nests showed a distinct peak in mid-June and another in early July, with a gap in late June. I didn't expect this. At this point, I could have offered arm-waving speculations to explain the observations. In a paper I published on this study, however, I explicitly avoided any conjectures about what might have caused this pattern.[7] When the time came to present the work at a scientific meeting, however, I was not so timid. I suggested that this pattern could reflect a tendency for female cowbirds in the area to produce eggs in two pulses—they might be inherently double-brooded. A senior and highly respected ornithologist in the audience, S. Charles Kendeigh, took exception to this interpretation. Of course, I defended my interpretation, rather belligerently I'm afraid.[8] My explanation was at best a hypothesis, although not one that could be tested without detailed laboratory and endocrinological studies. My interests were shifting elsewhere, and I didn't follow up (nor, to my knowledge, did anyone else).

The cowbird study started me on the pathway to becoming a *bona fide* scientist (of the ornithological variety). It provided an early exercise in dealing with unexpected results, a challenge that I would encounter often in my research career. My approach in the cowbird study was strictly observational, with no specific study design or sampling strategy. In fact, the description of methods in the publication of this work contained but a single sentence: "Nests in this area were tagged and observed three or four times per week until young fledged or the nest was deserted or destroyed."[9]

I knew, however, that doing science often entailed experiments, so in one of my class projects at the OU Biological Station I manipulated nature rather than simply observing and describing it. For some reason (I'm not sure why), I had become interested in how the movements of animals might affect their local distribution. If I were to release several individuals at a single point, what sort of distribution would be produced by their subsequent dispersal? I expected that most would move a short distance, but a few would go farther—a classical distance decay function.

At one of the farm ponds in the area I caught several dozen small frogs (spring peepers), marked them by clipping one toe, and released them one evening from a single point on the edge of the pond. I planned to systematically survey the margins of the pond at regular intervals over the next several days to determine where and how far the marked frogs moved. All carefully designed, I thought. That evening, however, an intense Oklahoma thunderstorm (a "gully washer") moved in. Several inches of rain filled the pond to overflowing. The next morning the shoreline where I had released the frogs was nowhere to be seen, submerged under several feet of water. None of the marked frogs was ever seen again. It was a rude introduction to the realities of conducting "experiments" in the field, and it probably strengthened my resolve to base my future work on observations alone (at least for a while).

By the end of my second summer at the OU Biological Station, the general outline of a career pathway had emerged. It started with my childhood *interests* in birds and my love of open spaces. These two passions set broad boundaries, within which my interests and studies would continue to grow and develop. This development was nurtured and reinforced by several key *people*. Bob Burns provided the encouragement that showed me a pathway into zoology, and then George Sutton opened the door. He took me into the field on class expeditions and collecting trips and hired me to help him on his magnum opus on Oklahoma birds. And Chuck Carpenter was my unofficial mentor during my summers at the OU Biological Station. But the other forces that can direct one along pathways had yet to come into play. The questions that I asked came from my interests and curiosity, not the scientific culture of ornithology or ecology. My studies had no detectable societal relevance and dealing with the practicalities of life still seemed years away. I was simply learning how to observe, gather data, and translate questions into a research project that might provide some answers.

By this time, I was determined to become an ornithologist. But capitalizing on motivation requires opportunities, and I benefited from several. My family's connection to George Sutton, going back decades, provided the catalyst to turn my childhood passion into a career pathway (although it required great patience on his part). Because I was the child of a faculty member, several zoology or botany professors at OU reinforced my teenage enthusiasm by including me on field trips with their students, which gave me a glimpse of what field research was all about. And the flexibility of the University College curriculum at OU provided the opportunity for me to probe various options while still graduating on time.

Such itinerant browsing is now much more difficult. There is pressure to declare a major field of study early in an undergraduate program, and then many courses are required to ensure that you are well-versed in the basics of a science. This leaves little time for exploring other areas that might interest you. Yet preparation for the next steps in becoming a scientist involves more than coursework. Experience in conducting research, either as an assistant on someone else's project or something of your own design (e.g., following the fates of nesting birds or releasing frogs to see where they go), may tell you right away whether a career as a research scientist is in the cards for you. Such extracurricular activities can weigh heavily when it comes time to apply to graduate programs to continue on your pathway to becoming an ecologist. More to the point, if you are not inclined to take advantage of opportunities to expand your training beyond a curriculum, you should probably ask yourself whether this is really the career pathway you want to follow. Does it align with what *really* interests you? If it doesn't, you should probably reconsider.

3

STARTING ON THE PATHWAY TO BECOMING A SCIENTIST

By the time I finished my senior year at Oklahoma I knew that I wanted to immerse myself in the world of birds. There could be many ways to do this but given my academic upbringing the obvious pathway led to graduate school. That way I could continue my training to become a certified scientist—an ornithologist (I wasn't yet thinking of becoming an ecologist). Although staying at OU was an option, I grew up in Norman. I needed to branch out, to experience other places, and learn from other teachers. It was time for a change.

Graduate studies are an apprenticeship to becoming a scientist. The choice of a graduate program is therefore critical in shaping you into a professional and revealing which pathways are there to be followed. But selecting a graduate program can be an arduous process. You must consider the faculty; potential research advisors; what interests other students pursue; other programs offered by the university; the community setting; and a host of other factors. But it's two-sided: the program and an advisor must also select you. You pick places and programs and send off applications, hoping some will bear fruit.

So I applied, but to only two places—the University of Arizona in Tucson and the University of Wisconsin-Madison. George Sutton suggested that I consider Arizona because there was a well-established program that emphasized the systematics and distribution of birds, which aligned with what I'd been doing at the OU museum and what I thought I was interested in going forward. There was an enthusiastic group of students. And I thought that living and doing research in the open spaces of the desert would be neat. Chuck Carpenter urged me to apply to Wisconsin, I think because he wanted to steer me toward studying animal behavior, which was his specialty. He'd also spent some time with John Emlen, who led the animal behavior program at Wisconsin. Chuck thought the two of us would be a good match. I knew something about the environment of southern Wisconsin from spending time there with my aunt and uncle when I was growing up. Mainly, I knew that the winters were harsh.

Location has a powerful effect on how you develop as a scientist. Arizona and Wisconsin would expose me to different environments, with different birds to see and watch (and collect). Interactions with different people would prompt different questions and pique different interests, leading me onto different pathways. If I went to Arizona, I'd be able to develop my skills in systematics and museum work. I could pursue distributional studies in exotic places like I'd seen some of Sutton's graduate students do. If I went to Wisconsin, I'd be joining a behavior group working at the forefront of the developing field of ethology, about which I knew little. In either case I could continue working with birds. I recall complaining to Sutton that Wisconsin didn't offer much in the way of systematics and distribution, so Arizona seemed to be a better choice. He counseled me to keep an open mind, that I'd find either to be enriching and I could always go elsewhere for a PhD degree if I wanted.

It turned out that my choice was determined by circumstances—serendipity. Wisconsin offered me a teaching assistantship that would provide financial support. Despite my entreaties, Arizona couldn't commit to similar support. A pathway had been chosen, although perhaps not for the most idealistic reason, and not by me.

I was off to Madison to join John Emlen's group. Emlen would be an ideal mentor. He had turned a childhood passion for birds into a successful career as a scientist. His PhD was in ornithology. His research, conducted in many places on a variety of species, emphasized the importance of observations, of studying animals where they lived, and of spending time out in nature.

Emlen had a large stable of graduate students, most of whom were studying animal behavior. Their approach was firmly rooted in the Tinbergen-Lorenz tradition of ethology, which emphasized how animals behaved in nature and the proximate mechanisms and ultimate evolutionary forces that determined behavioral patterns. No one seemed interested in the subtleties of subspecies differences or phylogenetic relationships of birds that had captured my attention at Oklahoma. But the excitement of the group was contagious. I soon found my interests shifting, inevitably, toward animal behavior. It was all new to me. I realized that I had much to learn.

At the same time, the realities of life began to affect my decisions. At the end of the year, I would marry Martha. Martha had been my high school sweetheart and we had continued dating off and on through college. When I left for Wisconsin, we decided we wanted to stay together. At that time, in the early 1960s, marriage was the most appropriate way to do that. When Martha came to Madison that winter, I would have new family responsibilities. I couldn't afford to dawdle; I needed to come up with a research project so I could move ahead on my master's degree.

Toward the end of my first fall at Wisconsin, Emlen and I sat down together and talked. And talked. Even though many of his students were involved in laboratory studies of the development of behavior in individuals, innate releasing mechanisms, imprinting, and the like, Emlen was at heart a field naturalist. He saw right away that my interests lay in the field rather than the lab, so we discussed possibilities. Somewhere along the way he mentioned that there was an interesting and unusual situation close at hand. A small cattail marsh on the shore of Lake Wingra (in the university arboretum) supported a breeding colony of red-winged blackbirds. That was not unusual; red-wings are common in marshes and wetlands throughout the Midwest. A previous Emlen student, Bob Nero, had studied the behavior of blackbirds in this marsh for his doctoral research several years before, so there was not a great deal more to be learned about their behavior. But another species, common grackles, had begun to establish nesting territories in the marsh, and there was now a thriving colony mixed in with the red-wings. Although grackles are common, and nest in a wide variety of habitats, they rarely do so in marshes. It might be interesting, Emlen suggested, to see how the species interacted with one another, what behavioral displays they used, and whether there were any effects of their interactions on their production of young.

I latched onto the suggestion. It would take me into the field. It would build on my nascent interest in species interactions, hatched in the cowbird study. It would involve finding nests and following their fates, as I had done in Oklahoma. And it would force me to look much more closely at the details of how animals behave and what comes of their behaviors. The field site was also close by, so it would be possible to balance the fieldwork with the demands of my teaching assistantship and my new family responsibilities. And it would be fun, I thought, to immerse myself in the world of marshes. I was shortly to understand the true meaning of "immerse."

EPISODE: A MARSH IN WISCONSIN

Marshes have their own beauty and mystique. On early spring mornings when the air is cold the mist rises from among the cattails and reeds. It's easy to see why people imagined spirits and ghosts lurking in the mists, like something from the writings of Tolkien. But as the morning sun breaks through, the marsh becomes a hotbed of activity. The abundance of emerging insects supports breeding warblers, wrens, and rails as well as blackbirds, and songs fill the air. Turtles emerge to bask in the midday sun.

The American ecologist Paul Errington wrote compellingly about the beauty and ecological importance of marshes in his book, *Of Men and Marshes*.[1] Over the past century, however, many marshes have been drained to make way for cornfields, soybeans, or housing developments. Marshes have become an endangered ecosystem and a focus of conservation efforts. My thinking about conservation, however, was still years in the future. For now, my focus was on birds.

The first step in undertaking a research project is to be clear about what questions you're asking. At this point, my questions were simply: How did the two species interact? Did the interactions have any consequences? To address your questions, however, you must figure out how to do so. How could I study and document the subtleties of blackbird and grackle behavior in a marsh? The birds nested among the cattails, which were dense and five or six feet tall. How could I even see the birds? And there were a couple of feet of water in the marsh, making it difficult to get around. I got some waders so I could go out into the marsh to map territories, find nests, and follow their fates. But moving through the marsh to observe behavior would disturb the birds and invalidate my observations—they would be responding to me rather than to one another.[2]

Bob Nero had faced the same challenge when he studied red-wing behavior in the marsh. His solution was to build a twenty-foot tower in the middle of the marsh, topped with a platform from which he could watch the birds. Sitting quietly on the tower, he blended into the marsh and the birds went about their business, presumably unconcerned about his presence. Nero's tower was still there, needing only a few minor repairs. I (literally) had a platform for my observations. I prepared to spend my days sitting atop a twenty-foot tower in a marsh.

Red-wings arrived on the marsh in early March. The silence of winter was now broken by the *o-ka-leeee* calls of the males as they began to establish territories. Activity picked up when the females began to arrive a week or two later, launching the males into an orgy of chasing females, posturing to one another, and defending territorial boundaries with ritualized displays.

Sitting on the tower was miserably cold. There wasn't much to see, as the grackles had yet to arrive. It was boring. But I didn't have long to wait. Male grackles showed up by mid-March, although they didn't begin to make forays into the hotbed of red-wing activity in the marsh itself until female grackles appeared a week later.

Time to begin serious observations. To do so, however, I needed to get to and from the tower in the middle of the marsh. When the red-wings and grackles arrived, Lake Wingra was still covered with ice and the marsh was frozen solid. With the first hints of spring, the ice in the marsh began to melt. But it did so from the top down, so there was a layer of ice hidden beneath the water on the surface. As more of this ice melted it did so unevenly, so there were gaps in the ice layer that were open down to the bottom of the marsh. More than once, I managed to find these gaps, plunging unexpectedly into deeper water that overflowed the tops of my waders, filling them with incredibly icy water. Looking back, I wonder whether this might have ordained my later attraction to warm, arid environments.

As the weather warmed and the red-wings and grackles began nesting, I learned to avoid the pitfalls in the marsh. I began recording observations of behavioral interactions, finding nests, and monitoring their success. I wanted to see whether red-wings defended their territories against incursions by grackles and vice versa, and whether the species used the same displays in their interspecific interactions as they did against one another. If there was considerable strife between the species it could affect their reproductive success, as it might take time and energy away from normal reproductive activities such as feeding young. If extreme, this strife could lead to a neglect of their own offspring (what ethologists labeled "aggressive neglect").

Grackles were clearly the interlopers in the marsh. Most grackle nests were located within red-wing territories, prompting aggressive responses from the red-wings. When grackles approached their nests to feed chicks, they often skulked low through the cattails, sneaking to the nest and quickly flying away once the young were fed. In contrast, red-wing females (only females fed the young) flew directly to their nests. They made no obvious efforts to remain inconspicuous and were ignored by the grackles. When grackle chicks left the nest, the adults quickly led them out of the marsh, often by enticing them with food and then moving a short distance away before actually feeding them. As grackles moved through the marsh with their young, red-wings repeatedly dived at the grackles, chasing them from their territories. Grackles with nests toward the center of the marsh therefore had to run a gauntlet of territories, eliciting a cascading series of attacks.

I spent two breeding seasons sitting on that tower in the Lake Wingra marsh, recording and quantifying the behavioral interactions between the species. But what about the bottom line: Did the interactions affect the reproductive success of either species? Not so far as I could tell. Roughly 50 to 60 percent of the eggs laid by both red-wings and grackles produced fledglings, close to the range of nesting success I found in other "pure" colonies of red-wings and grackles in the Madison area. By this measure, then, the behavioral interactions between the species were inconsequential.

At the end of the study, I had documented the nature and extent of behavioral interactions between red-wings and grackles and assessed their effects. The questions I asked were, essentially, who did what and did it matter? Nonetheless, when it came time to write up my findings, I cast my observations in the behavioral jargon of the time—such things as "approach-withdrawal conflicts," "fixed action patterns," "social releasing mechanisms," "sign stimuli," "habituation," and "interspecific territorialism."[3] After all, I was now surrounded by a bunch of ethologists and I wanted to fit in. In retrospect, I was probably trying to wedge simple behaviors into complex conceptual niches with sciency names.

What had begun as a descriptive study ended that way. Over the years I had progressed from watching birds and finding nests to collecting, and then to making observations as part of a *bona fide* research program. But I was not guided by any theory or overarching paradigm, nor did I test an *a priori* hypothesis. I simply recorded observations of an unusual situation to find out how the two species got along with each other. In the process, I learned how to conduct field research—the importance of careful observation, quantifying what I saw, and patience. I was moving beyond the coursework that is the foundation of a scientist's formal training to the next steps of asking questions and formulating research to answer them. I was now on the pathway to becoming a scientist.

I had gotten to this point because of the forces and factors that I highlighted in the Introduction. My *interest* in birds defined what I would study, and *motivation* dictated that I would study them. *Opportunity* determined where I would study them. *People*, from George Sutton to John Emlen, provided the encouragement that led me to sit on a tower in a marsh for two summers. And the realities of my *personal life* kept me in Madison instead of gallivanting to far-off places. A different combination of forces and factors would have directed me toward a different pathway, although the driving force of my interests would have ensured that it would involve birds and follow the formal steps to becoming a scientist. The other factors that can determine one's pathway—societal relevance and the culture of a science—had yet to come into play.

4

DEFINING A PATHWAY

My interest in studying birds was unabated when I finished my observations of blackbird-grackle interactions. Although my assistantship at the university now included preparing study skins and curating the bird collection in the UW Zoological Museum (UWZM), my earlier determination to focus on systematics and distribution had waned. I was now thoroughly captivated by animal behavior, a consequence of my decision to go to Madison instead of Tucson. But I was less interested in the mechanisms of how and why animals behaved as they did—the domain of classical ethology—than in how behavior affected their relations with one another and with their environment—behavioral ecology. John Emlen's own work emphasized the linkage between behavioral mechanisms and their ecological context, and his graduate course was called "Behavioral Ecology." I decided to emphasize behavioral ecology in my PhD.

This is the time in your graduate training when you need to be thinking about whether established scientists will find your questions and research interesting and important. This will affect your likelihood of getting a job, obtaining research funding, and publishing papers—all the things that are necessary if you are to join the scientific culture when you complete your degree. As I contemplated continuing graduate work toward a PhD, my interests turned toward the behavioral ecology of species interactions.

At this time (1963), concepts and theories about species interactions and ecological communities were in active ferment. G. Evelyn Hutchinson and his student Robert MacArthur were bringing new theoretical and mathematical rigor to studies of differences in habitat or resource use among species—"niche partitioning"—that had been a focus of ecological thinking for decades. Ecologists were emphasizing the role of competitive interactions among species in producing those differences and determining which species could coexist in ecological communities. In his own PhD study, MacArthur had used observations of the foraging behavior of five warbler species to show the subtle ways in which they partitioned resources to avoid competition.[1] Such niche-partitioning studies were the foundation of an emerging view that competition among ecologically similar species was a (or arguably *the*) major force determining the structure and species composition of many animal communities.

Here was a conceptual foundation that I could use to generate some answerable questions and frame my work in a way that would resonate with my future ecological peers. How did ecologically similar species interact behaviorally? Which species were dominant to others? How did their habitats differ? Did the patterns agree with the predictions of theory? The blackbird-grackle study had emphasized the behavioral side of species interactions; could I now expand this approach to incorporate the ecological context and consequences as well?

Where would I find a suitable assemblage of species to study? I could continue to dig into the details of interactions between redwings and grackles, perhaps by documenting differences in the food items they brought to their nestlings or where and how they obtained them. But this would involve only two species, and I was ready to leave the marsh behind. I wanted to try something different. John Emlen had spent some time in Africa studying closely related species of weaver finches that occur together in the same habitat, so that was a possibility. Or there was an opportunity to work with seabirds on the Pribilof Islands in Alaska, where multiple species nested together and foraged out at sea. Where and on what I decided to conduct my research would play a big part in determining my future pathway.

Once again, however, life's practicalities intervened. By now Martha and I had a baby daughter, Ann. The thought of being gone for months at a time, missing precious moments, was unbearable. Ann was curious about the world, as most young children are. I felt I had a responsibility to make sure she experienced and learned to appreciate nature. So I began to look into what I could do locally.

John Emlen and I talked some more. And talked. He mentioned that another of his previous students, Wesley ("Bud") Lanyon, had studied two meadowlark species at Fitchburg, just south of Madison. Lanyon found that the meadowlark species excluded each other from their territories as forcefully as they excluded others of their own species. This was the sort of behavior that aligned with my interest in species interactions. Five other bird species also nested there. Documenting the behavioral and ecological interactions among these species would be interesting but not overwhelming. Although the plant species composition of the Fitchburg study area was not that of a native prairie, it was structurally a grassland, which fit with my childhood attraction to open spaces. I would be able to see how individuals related to their habitat and to one another, without spending days sitting on a tower.

"Grassland," however, is only a general description of habitat. MacArthur's warbler study had shown that a much more detailed and nuanced assessment of habitat features would be needed to tease apart the niche differences among the species. I would have to develop quantitative measures of "habitat" and design a sampling approach that would permit statistical analyses. Emlen had recognized the importance of habitat description and measurement in his work with weaver finches in Africa. He was now keen to see what I could do with grassland birds in Wisconsin.

EPISODE: BIRDS IN A WISCONSIN GRASSLAND

Developing a methodology for assessing habitat relationships among species was to become a major focus of my dissertation research. Where to begin? All of the species in the study area defended well-defined breeding territories. Once an individual had established its territory it stayed within the boundaries. The key, I felt, would be to relate the behavior of an individual to the habitat available in its territory. To do this I would first have to determine the territorial boundaries. I found that individuals would stay within their territories as I chased them around, so I could map territories by connecting the dots where they flew and landed (what I came to call the "territory flush" method).

These mapped territories became the focus of my habitat analyses. At sampling points within each territory, I measured multiple features of the habitat that I thought might be important to the birds—over thirty-five habitat variables in all. These measures allowed me to quantify the habitat features of areas occupied by each species on a territory-by-territory basis. But I also wanted to know how the birds used their habitats. For this, I recorded timed observations of the behavior of individuals and the habitat characteristics where specific activities (such as singing, foraging, or aggression) occurred. Unwilling to give up my compulsion for finding nests, I measured habitat features of nest sites and diligently followed the fates of the nests.

This inordinate attention to sampling design and habitat measurement stood in stark contrast to my earlier cowbird study, or even the red-wing-grackle research. Those earlier studies were entirely observational: find nests, watch birds, and record what I saw. Conducting the research required of a PhD, however, demanded greater attention to methodology and quantitative detail. But it also reflected John Emlen's long interest in rigorous observational methods. Rather than being a slave to what textbooks prescribed, he would let the behavior of the study organisms (usually birds) tell him what was important and then develop innovative ways to systematize his observations. For me, this was the beginning of a long-standing interest in adapting how observations and data are gathered to the individuals or organisms being studied.

I spent the summer of 1964 getting to know the landowners, becoming familiar with the study area and the birds that nested there, and developing and refining methods of habitat measurement and description. That fall, at John Emlen's invitation, Robert MacArthur came to the UW Zoology Department to talk about competition and community structure. MacArthur was now leading the way in integrating population biology, evolution, and the geographical distribution of plants and animals (biogeography) into a cohesive body of mathematically based theory in ecology. But he was also an inveterate field naturalist with a particular interest in birds. I was eager to see what he thought of my approach, so I signed up to meet with him. I excitedly (I'm sure) described my study plan. He thought it was interesting and novel, although perhaps overly detailed. Along the way, MacArthur mentioned that one of his students, Martin Cody, was beginning similar dissertation research. It might be good, he suggested, for the two of us to get in touch.

The next week I sent MacArthur a lengthy description of what I was doing, which he shared with Martin. Martin replied, outlining his project: "I hope to obtain some set of parameters applicable to all grasslands which may be important in the habitat selection of grassland birds, and use the set to compare the strategies of various bird communities in avoiding competition."[2] This was pretty much what I had in mind. I was crestfallen. I could see that the two of us, both studying competition among grassland birds, might end up being competitors ourselves. We briefly considered adjusting our approaches to reduce overlap—partitioning our research niches, so to speak. Neither of us wanted to change our plans, however, so we rationalized that we were taking different approaches and working with different species in different places, so our projects were likely to be complementary rather than competitive. We continued on our separate ways.

What did the methodology and measurements that I developed produce over the three years of the study? Although I obtained lots of detailed information about the ecology and natural history of each of the seven breeding bird species, my primary interest was in the similarities and differences among the species in the habitats they occupied, how they used the habitat, and how they interacted with one another. I documented territory establishment, sizes, and overlap between species. I measured an array of habitat features at sampling points within territories and at nest sites. I spent many hours recording the behavior and activities of individuals to determine how each species spent its time. I tabulated the habitat features associated with each activity. I tallied behavioral interactions among the species to determine who dominated whom.

By the time I finished fieldwork and was writing my dissertation, the "competition paradigm" had become well-established in ecology. In this view, competitive interactions among species would lead to the exclusion of marginally adapted species from communities, leaving assemblages in which the remaining species were optimally separated in their use of resources such as habitat and food. It was a "Goldilocks" solution: coexisting species would not be so similar that competition would reduce their ability to reproduce, nor so different that important resources would be left unused. Rather, differences among species would be "just right" to enable them to share resources while getting on with their business. In the fashion of Kuhnian normal science, ecologists were busy documenting niche differences among ecologically similar species and interpreting those differences as evidence of the role of competition in structuring communities. The paradigm provided an attractive framework in which to cast my conclusions, one that would resonate with the broader community of ecologists that I now intended to join and appeal to them when the time came for me to submit proposals for funding to continue my research.

As I analyzed my observations and synthesized the results, however, it became apparent that similarities among the species were much greater than the differences. Although there were slight differences in habitat features such as overall vegetation height, density, and litter depth among territories of the different species, structural features of habitats were broadly shared. There were behavioral differences in how the birds used habitat features, although they also were not great. Nonetheless, the power of the competition paradigm prevailed. I concluded that "differences in habitat occupancy and utilization were seemingly adequate to circumvent direct competition and to explain the co-occupancy of the study area by the seven breeding species."[3]

～⌒

What about Martin Cody? We both completed our PhD studies at the same time. As we anticipated, our approaches to studying grassland bird communities had diverged. Like me, Martin focused on vegetation structure and foraging behavior as the primary ways in which species might differ. But while I studied one large area intensively to document habitat occupancy and use by seven species throughout the breeding season, Martin spent less time on each of ten, widely scattered, smaller study plots in North and South America that each contained three to four breeding species. My approach emphasized the methodology of data collection; Martin excelled at innovative ways of analyzing the data once they were collected. We both concluded that our observations were consistent with the expectations of competition theory, although he went further to state that his sites contained stable sets of species that were optimally separated ecologically.

My probing of the results of my Wisconsin study didn't end there, however. I had initially assessed the similarities and differences among the species at the peak of the breeding season, when the birds were raising young, demands on resources were likely to be greatest, and potential competition most intense. At this time, the territories of most species broadly overlapped and blanketed most of the study area, so it wasn't surprising that differences in the habitats within their territories were slight. A few years later, however, I started wondering whether differences among the species might be greater among the first birds to arrive and establish territories in the spring. I had mapped territories at regular intervals throughout the breeding season, so I could ask whether there were habitat differences between the first territories established and those occupied later, at the peak of the breeding season.[4]

There were enough territories to address this question only for savannah sparrows and grasshopper sparrows. My previous analysis had shown only slight differences in habitat features of territories of these two species at the peak of the breeding season. Yet when they first arrived in the spring the two species occupied significantly different portions of the range of habitats available to them. The earliest savannah sparrows settled on areas that by mid-June would have deeper and patchier litter, greater grass cover, and less coverage and density of shorter forbs. Grasshopper sparrows showed opposite relationships. As more individuals arrived and the study area filled up, however, overall habitat occupancy by the two species converged. Habitat niche partitioning between the species was a matter of when one looked. My earlier conclusions about competition and niche separation among the species did not seem so secure.

Graduate school is about much more than doing the research for an advanced degree. For me it was a place to form lasting friendships, relax with my compatriots over beers and brats, and carry on deep discussions of important and not-so-important issues of the day. Emlen's group of graduate students often got together for lunch to share our thoughts about goings-on in behavior and ecology—recent papers, books, meetings, and the like. This was where I got started on a side project that would end up combining my interests in behavior and evolution with my earlier attraction to the humanities and writing.

EPISODE: DELVING INTO GROUP SELECTION

In 1962, the British zoologist V. C. Wynne-Edwards published a formidable 653-page book, *Animal Dispersion in Relation to Social Behaviour*.[5] The issue of whether and how populations were regulated so they would not grow beyond the limits of resources had been a topic of considerable debate among ecologists for decades (indeed, since the time of Malthus).[6] Wynne-Edwards's thesis was that many forms of social organization and behavior had evolved to avoid overexploitation of food resources and regulate population size by shifting direct competition over food to "conventional rewards" such as holding a territory or a position in a dominance hierarchy. So-called "epideictic displays" among individuals could be used to gauge the density of a population. In response, individuals would adjust their habitat occupancy, resource use, or reproduction accordingly to prevent the population from overexploiting its resources. Well-regulated groups would persist while those that lacked the appropriate social behaviors would perish. The interests of the social group would override those of its individual members. Consequently, natural selection would operate at the level of social groups and supersede selection based on individual self-interests. In a nutshell, social behavior evolved through selection on entire groups—group selection— in order to regulate populations.

Wynne-Edwards brought many forms of social behavior and organization under the umbrella of his theory and supported his arguments with examples from plankton to primitive humans. There was plenty of fodder for our lunchtime discussions. Most of Emlen's students were working on behavioral projects, so our attention focused on Wynne-Edwards's treatment of social behavior. We agreed with some of his arguments and disagreed (sometimes vehemently) with others. By and large, however, we thought his theory represented an important effort to bring some conceptual cohesion to seemingly disparate forms of social behavior. It deserved to be taken seriously.

One evening over beers at a science conference, several of us from Emlen's group got together with some graduate students from other universities. The conversation drifted to Wynne-Edwards's book. Their reaction was intense: the entire theory was wrong and misleading, not the least because of its reliance on group selection (which we had not really discussed). Several prominent evolutionary biologists and population geneticists had expressed similar reactions. There was a real possibility that the important points in Wynne-Edwards's theory would be disregarded because of the controversial evolutionary mechanism that he proposed. I decided to marshal my thoughts into a paper.

To present a convincing argument, I'd need to address the issue of group selection head on. It wasn't something I knew much about. After spending several months digging into the history and thinking about group selection, I felt ready to discuss it with someone who *really* knew the topic. Years before, the evolutionary geneticist Sewell Wright had laid the theoretical foundation for thinking about levels of selection. He was now on the genetics faculty at Wisconsin. I asked him to read my draft critique, we talked at length, and he invited me to present my ideas to the graduate seminar he led with James Crow, another evolutionary geneticist. Their feedback clarified my thinking and assured me that I was not hopelessly naïve. I continued writing. After reviewing the nuances of group selection and delving into the details of how selection acting on individuals could produce adaptive social behaviors, I concluded that arguments about Wynne-Edwards's ideas should focus on the role of social behavior in population regulation and avoid getting distracted by his evolutionary speculations.

Why did I do this? Why would I take time away from analyzing my grassland bird data and writing my dissertation to pursue a side project that might not lead anywhere and was certainly not central to the pathway I was developing? I can only guess that it was because my interest was aroused, and I found it intellectually challenging to explore the concepts that Wynne-Edwards was advancing. My critique, "*On Group Selection and Wynne-Edwards' Hypothesis*," was published in *American Scientist* in 1966, shortly after I received my PhD.[7] The paper was unlike anything I had written before or would write for another decade. It dealt with *concepts* rather than data. Looking at it now, I realize that it reads more like the literary criticism papers I read as an undergraduate than a typical scientific paper. I referenced papers in a variety of disciplines and evaluated arguments and counterarguments. I used footnotes (which were forbidden in most scientific publications). My old attraction to the humanities hadn't disappeared, after all. I would write about concepts and synthesize ideas more often as I progressed professionally, no matter what pathway I was on.

The episodes that punctuated my graduate studies—making behavioral observations of red-wings and grackles, developing methods to document niche partitioning among grassland birds, and even dabbling with group selection—were nodes along a developing pathway. I had started as a college freshman uncertain (and largely unconcerned) about what to do and had now arrived as a freshly minted doctor of philosophy, eager to launch a career as a scientist. The quantitative rigor of my work had increased. My first studies had been strictly observational. I had paid scant attention to sampling and my methods involved little more than finding nests, watching birds, and taking notes. By the time I began my study of a grassland bird community, my attention to methodological details and sampling design had become almost obsessive.

I also learned how to ask questions and relate the questions to observations. Initially, I had no underlying questions; I just observed. Then questions about species interactions set the stage for my study of red-wings and grackles. As my focus shifted to a community of several co-occurring species, my questions now came from the competition paradigm; theory and the scientific culture determined what should be observed and what to expect.

My move from Oklahoma to Wisconsin had led to an abrupt redirection in my pathway. It was more than a change in universities, or where I would call home. The woods, wetlands, and agricultural fields of the upper Midwest replaced the varied habitats I'd come to know in Oklahoma. Many of the birds were different. I became part of a group whose interests were centered on animal behavior. My desire to forge a career in avian systematics and distribution had fallen by the wayside as I followed a trajectory toward animal behavior and behavioral ecology. And John Emlen provided a role model of how one can combine a passion for fieldwork with science that cuts across disciplines—in this case, behavior and ecology. I now knew where I wanted to head. My future pathway seemed clear.

There's a broader message here. Let your *interests* be your guide. Becoming an ecologist is not a straight-line journey but something that shifts as your interests evolve. Although the core of your interests may be enduring, how and on what those interests are expressed may change. And don't be afraid to follow side paths if your interests lead you there. Even if they don't determine your future pathways they may be intellectually rewarding and enhance your professional development. My paper on group selection was more widely cited than the monograph reporting on my dissertation research, probably because it dealt with concepts rather than data. The paper was even included in an anthology dealing with population regulation.[8]

5

BEGINNING AN
ACADEMIC CAREER

Completing a PhD is a lengthy process. But it is also exciting and, in its own way, comfortable. Interactions with fellow students and professors are intellectually stimulating and challenging. It's a sheltered way of life. It may be the one time in a professional career when you can single-mindedly pursue an interest or question, largely unfettered by the realities of a job that will come perhaps all too soon. It's easy to settle in and prolong a program—after all, one more experiment or one more field season will surely help resolve lingering uncertainties, tie up loose ends, or explore new ideas that may be a foundation for starting a research program when you enter the outside world.

But sooner or later a dissertation must be finished. It's time to graduate and move on, to explore new options. Do you pursue a postdoctoral appointment to continue your research, develop new skills, deepen your experience, or branch out in a new direction while still sheltered from the accountability that comes with a job? Or perhaps you just want to delay that fateful step of getting an actual job a while longer? And what kind of job should you seek? Should you consider a job in the private sector, perhaps with a consulting firm where your work would be guided by what the clients want? Or with a government agency, where you might be able to concentrate on conducting research, albeit on issues tied to an agency's mission and constrained by politics? Or do you search for a tenure-track faculty position in a college or university where you can combine research with teaching, guide graduate students, and serve on seemingly endless committees? The decision about what to do and where to do it marks a major intersection in diverging career pathways and determines what will follow.

In my case, it was a foregone conclusion that I would aim for an academic position. After all, I'd grown up in an academic setting with professorial parents and a deep family legacy of teaching. What else could I do? I also had family responsibilities to think of. Our daughter Ann would shortly begin kindergarten and then primary school, and it was important to provide her with some degree of stability. I'd spent five years in graduate school. Prolonging it further didn't make sense. I was anxious to move ahead. It was time to get a faculty job.

So I set about applying for faculty positions at perhaps a dozen universities. Competition for faculty positions was intense even then (although not nearly as intense as it is now). I did all those things that you do to make your application stand out. You try to match your qualifications as closely as possible to what's specified in a position announcement (I portrayed myself as an ornithologist, ecologist, behavioral ecologist, or ethologist). You emphasize how your research addresses prevailing issues that would be likely to attract funding (e.g., the role of competition in structuring ecological communities). You highlight your diverse experience (e.g., writing a paper on group selection). You cultivate well-connected experts who will write letters of recommendation for you (e.g., a member of the National Academy of Sciences who served on my graduate committee). And you publish early and often.

I was invited to interview at several places, but the ones I remember most clearly were Dartmouth College and Oregon State University. The two prospects could scarcely have been more different.

Dartmouth was a small private Ivy League university on the East Coast. The landscape setting was quintessential New England: patches of deciduous forest interspersed among small towns and farms. Oregon State was a public land-grant university on the West Coast less than fifty miles from the Pacific Ocean. Coniferous forests blanketed the surrounding mountains. Oregon State had perhaps four times as many students as Dartmouth. Undergraduate classes (which new faculty usually teach) would be correspondingly larger, with fewer opportunities for interactions with individual students. Research opportunities and teaching responsibilities at the two institutions would be quite different.

Where I ended up would establish the direction and context of my career pathway going forward. If I went to Dartmouth, I could initiate new studies of bird communities in northeastern deciduous forests or find the equivalent of grasslands in New Hampshire to continue beyond my doctoral research. If I went to Oregon State, I could study bird communities in the more extensive coniferous forests of the northwest and perhaps extend my grassland studies to remnants of the Palouse prairies in Washington. At this point, however, I wasn't thinking about alternative pathways and where they might lead. I just wanted to get a good job in an interesting place so I could continue investigating the behavior and ecology of birds and their interactions. Location would determine all else that followed.

I found the prospect of joining the faculty at Dartmouth most attractive, perhaps because it was a more prestigious institution, perhaps because when I visited the campus spring was in the air. If Dartmouth had offered me a faculty position, I would have taken it immediately. But they didn't make an offer (in fact, they didn't fill the position that year). Instead, Oregon State offered me a position.

Oregon State (OSU) certainly provided opportunities that aligned with my developing interests. The OSU Zoology Department that I would join had a rich tradition in natural history (replete with a small museum). The land-grant mission of the university ensured that there would be faculty colleagues and scientists with complementary expertise in areas such as botany, statistics, wildlife management, range science, and oceanography. The Pacific Northwest and its environments would be new to me, potentially opening new research opportunities. I'd be able to build a program in behavioral ecology by attracting graduate students. Corvallis, where the university was located, seemed like it would be a fine place to raise a family. And there was even a family connection of sorts: my uncle's first faculty position after he obtained his PhD in English was at Oregon State (although he was only there for one year). I accepted their offer.

We moved to Corvallis in the fall of 1966. I began teaching (Introductory Zoology, Ecology, and Ornithology), serving on committees, and learning the ropes of departmental and university politics as all new faculty must do. Three graduate students whose projects were already underway had been waiting for the arrival of the new ecologist/ethologist/ornithologist, so I became their faculty advisor. I also needed to establish a research program. I no longer had George Sutton, Charles Carpenter, or John Emlen at hand ready to suggest interesting research possibilities. I would have to come up with something on my own.

When I arrived at Oregon State, a new building to house the Zoology Department was under construction. When we moved in the following year, I discovered that the laboratory space I was assigned had been designed for "an ecologist," apparently with the expectation that the individual who was hired (i.e., me) would work in aquatic ecosystems. The lab was outfitted with several well-designed water tables for holding aquatic insects, fish, or amphibians. Instead, they had hired a bird guy who worked in grasslands. What could I do that would make use of the water tables?

EPISODE: TADPOLES IN THE LAB

The other vertebrate zoologist in the department, Bob Storm, wanted a colleague and he wanted me to succeed. He offered a solution. "Why not do something with frogs or tadpoles," he suggested.

"They're easy to maintain, and I could produce tadpoles for you on demand."

At that time, the idea that experience with a particular habitat early in life could determine an individual's subsequent habitat preferences—habitat imprinting—was a hot topic in behavioral ecology. Habitat imprinting had been investigated in mammals, birds, and fish, but not yet in amphibians. I could pursue my interests in habitat selection and behavioral ecology using the facilities I had been given, and I could do so during the rainy winter when I preferred to stay inside. So I readied the water tables while Bob produced the tadpoles. He injected a female red-legged frog (*Rana aurora*) with a pituitary suspension to induce ovulation, and when eggs were produced, they were fertilized with *R. aurora* sperm *in vitro*. When the eggs hatched the tadpoles were kept in featureless pans until I was ready to begin the experiment.

The design was simple. I set up two rearing "habitats" in the water tables. One had black stripes against the green background of the water table; the other had black squares. A third group of tadpoles, raised in a featureless water table, served as controls. After two weeks in their rearing habitats, tadpoles from each group were tested individually. I released a tadpole in the center of a test chamber where it had a choice between the half having black stripes or the half with black squares, both identical to the patterns in the rearing habitats. I wanted to see if their prior experience with striped or square patterns influenced their choice. If early experience had an effect, those raised in the striped habitat would prefer the striped half and those raised in squares would select the square-patterned half. Alternatively, if early experience had no effect they would, like the control tadpoles, show no preference. I had a clear hypothesis to test, with clear predictions.

I tested several dozen tadpoles from each group. The control tadpoles, as expected, showed no preference for stripes or squares. Neither did the tadpoles raised in the square-patterned habitat. But those raised in the striped habitat showed a strong preference for the striped half of the test chamber. This preference was acquired rapidly and persisted even after the tadpoles were kept isolated in a featureless tank for several weeks after testing, and then tested again. Moreover, the preference for stripes persisted through metamorphosis as the tadpoles became juvenile froglets.[1]

These results were perplexing and completely unexpected. I began to think that I had botched the experimental design. When I released the tadpoles in the choice chamber, for example, I had hovered overhead to record their responses. Perhaps I somehow influenced their choices (but why then only the stripe-reared tadpoles?). I decided that I needed to pay closer attention to how the experiment was conducted. So I repeated the experiment the next year using the same design but with more rigorous procedures. Same results. Tadpoles raised in the stripe-patterned habitat selected stripes when tested; the other groups showed no preferences. The results were not artifacts of my experimental design; they were real. It was time for some arm-waving speculation (the sort of thing ecologists are good at doing).

My decision to use stripes and squares for the habitat patterns had no a priori biological rationale. I had simply made the stripes and squares out of black electrical tape that I had at hand, reasoning that this presented bold, contrasting patterns that should be easy for the tadpoles to learn. But why did tadpoles respond to a striped pattern but not squares? Here's where the "ecology" in behavioral ecology came in. Rana aurora breeds in shallow ponds or overflow areas that contain submerged weeds or grasses, filamentous algae, cattails, and slender willow branches—all linear structures that would cast linear shadows on the muddy substrate. This habitat is often patchy, areas of submerged and emergent vegetation interspersed with open areas that have a uniform substrate. Tadpoles tend to stay within their natal patches, but as they grow and become active, they move about a good deal. Perhaps by forming an early preference for habitats with linear patterns (i.e., stripes) tadpoles would be likely to orient to such areas in a patchy environment and then remain in a habitat that might afford more protection from predators. Conversely, by not developing a preference for other habitat patches (i.e., square-patterned or uniform) they would continue moving until they encountered a more suitable (i.e., striped) habitat patch, and then they would stay there. This was a *post hoc* explanation, to be sure. But it could also be framed as a testable hypothesis. Tadpoles of another species that bred in an equivalent of the square-patterned habitat might be expected to develop a preference for squares but not stripes. Fortunately, Bob knew of another closely related species, the cascade frog (*Rana cascadae*), that breeds in clear mountain ponds that often have a gravel (i.e., square-patterned) substrate but generally lack linear objects. The following year I repeated the experiment using *R. cascadae* tadpoles that Bob produced. Everything else remained the same. The control tadpoles and those raised in stripe-patterned conditions showed no preference for stripes or squares. The tadpoles raised in square-patterned habitats, however, displayed an overwhelming preference for the square-patterned half of the test chamber.[2] My initial hypothesis seemed to hold up. On the basis of their early experience, tadpoles of the two species selected stylized habitats that were consistent with their breeding ecology.

~❦

Flushed with the success of these experiments and enamored with my clever explanation, I was tempted to continue investigating anuran habitat selection. Several of the forces that I mentioned in the Introduction had pushed me in this direction: *people* (Bob Storm) had played a critical role, and the research topic (habitat imprinting) was currently being promoted by the *scientific culture*. But the tadpole studies had been initiated because of the physical facilities available to me, not because of a deep burning interest. The work was in the laboratory with amphibians, not out in nature with birds where I wanted to be. Open spaces beckoned. So I abandoned that pathway. That was a road not taken. The take-home message: follow your interests.

6

EXPANDING MY VIEW OF GRASSLAND BIRDS

Shortly after I arrived at Oregon State, I realized that I needed to submit a grant proposal to the National Science Foundation (NSF) to fund my field research. And I needed to do it quickly so I could begin in the next field season.

Submitting your first grant proposal is one of the most critical steps for a freshly minted PhD or new faculty member. If funded, it will determine your primary research pathway for the next few years. Reviewers are generally willing to cut new investigators some slack, but there are several points that may enhance your probability of success. Your proposal should be innovative but not too ambitious. It should offer new insights or address an issue of current interest, something that will excite your peers. It should fit with the priorities of the granting program. To demonstrate your ability to do what you say you will, it should build on what you have already done. And it should be feasible—doable.

I doubt that I considered all these points when the time came to submit my first grant proposal. I hadn't given up on the idea of working with birds and grasslands. The tadpole experiments, after all, were conducted during the winter when the birds that I would study would be off on their wintering grounds. So I fell back on what I knew best. The Wisconsin study had been designed to develop a methodology for describing and measuring habitats of grassland birds and then testing it with an intensive study at a single location. My conclusions about competition and community structuring developed only along the way, as the competition paradigm gained force in ecology. Now, however, I could use the conceptual framework of this paradigm to compare the structure and organization of grassland bird communities over several study sites spanning a range of conditions, rather like Martin Cody had done. I would use the habitat-description methodology that I developed in Wisconsin to document the details of niche partitioning among grassland birds.

I submitted a proposal to NSF, boringly titled: "Aspects of Avian Ecology in Grasslands." I proposed to conduct surveys of breeding birds and measure habitat structure at grassland sites ranging from tallgrass prairies in Kansas, to shortgrass prairies in Colorado and Wyoming, to montane grasslands in Montana, and Palouse prairie in Washington. To appeal to the broader community of ecologists, I dressed the proposal in the cloak of the prevailing competition paradigm: animal communities are in an equilibrium determined by competition among species to reduce niche overlap. I would document niche differences among birds in a variety of grasslands to see whether the patterns that emerged were consistent with the predictions of the paradigm. That is what I told NSF. In fact, I had no expectations of what I might find or specific hypotheses to test. I simply wanted to make comparisons to see what turned up. And I wanted to spend more time in the grasslands that I loved.

Despite the lack of specific hypotheses, NSF approved the proposal the following spring—for three years of funding. If my proposal had not been funded, I might have dug deeper into tadpole habitat selection. More likely, I would have focused on some aspect of bird communities and habitats in the nearby coniferous forests, perhaps drawing on MacArthur's work relating bird species diversity to foliage height diversity in eastern deciduous forests.[1] The methods of habitat measurement that I had developed in short-stature grasslands, however, were not designed to work in tall, multilayered forests. I would need to devise a different approach, perhaps by modifying MacArthur's methods. But NSF's decision had determined that my pathway would continue through grasslands, at least for a while.

Here also, MacArthur had a role to play. Because it drew on his theoretical work, he was asked to review my proposal. He gave it high marks (no doubt contributing to its success) and followed up with a letter. MacArthur wondered whether my approach, while suitable for an intensive study of the birds on a single site, might be too detailed to be useful in a comparative study. He was right, of course. I planned to visit multiple sites at the peak of the breeding season, so I could spend only a few days at a site before moving on. Making detailed observations of birds' behavior, activities, and nesting on each study site, as I had done in Wisconsin, would not be possible. I decided to restrict my activities to mapping territories and measuring habitat features. MacArthur also saw that I had not framed an *a priori* hypothesis. As he put it, "what kind of results should the comparison come up with?"[2] He confessed that he often asked the same question of his own studies.

Because I didn't know what to expect or have an explicit hypothesis to test, I wasn't sure what to look for. My questions were simply, how did the birds on a site differ in habitat occupancy, and how did the patterns vary among sites? I would gather data, compare patterns among sites that differed in vegetation structure, annual precipitation, and grassland type, and then interpret the results in the context of competition theory, as I had done in the Wisconsin study.

EPISODE: LOOKING FOR COMPETITION IN GRASSLANDS

Figuring that a bare-bones budget would increase my likelihood of success, I did not include funding for students or assistants in the proposal to NSF—I intended to do all the fieldwork myself. For the next three summers I spent most of June and July touring grasslands from Kansas to Washington while Martha and our kids visited her parents in Oklahoma.

The first challenge was to locate suitable study sites: grassland areas large enough to contain a twenty-five-acre (10-ha) plot with a sizeable buffer of surrounding grassland. Most sites ended up being on research areas managed by universities or federal agencies, although a few were on private ranches. At each site I laid out a grid of reference points marked by tall wooden stakes, mapped territories of the bird species present using the territory-flush technique I described earlier, and measured habitat attributes at sampling points using the methods I had developed in Wisconsin. I stayed on the study sites, sleeping in the back of a station wagon but going into nearby towns for dinners (and, since it was hot, a beer or two). Most of the sites were grazed, so sometimes I had to contend with cattle moving through my plot, eating the plastic flagging I tied to grid stakes or knocking down the stakes by scratching themselves.

Local landowners were sometimes interested but always a bit mystified by why someone would show up in the middle of a field in the heat of summer, pound a bunch of stakes into the ground, and then chase birds around for a few days before pulling up stakes and disappearing. My encounter with a rancher in Kansas was typical. After watching me from his pickup on the roadside for a while, he got out, ambled over, and asked, "Hey, whatcha doin' here?" I explained that I was from Oregon State University and was conducting research to find out how birds used grassland habitats.

"Oh, you mean those dickey birds? I didn't know anybody cared about them."

As soon as I launched into an explanation of the roles that birds could play in the functioning of grassland ecosystems, I could see his eyes begin to narrow.

"You ain't here to mess with my cattle or tell me where I can or can't graze, are you?"

"No, no, I'm not with the government," I said. "I'm just interested in grassland birds. Have been since I was a kid." "And this is really a special place you have here," I added, "especially around sunset."

This seemed to please him, although it was obvious that he thought me a bit crazed by the heat. "Well, you be careful. I saw a big rattler here just the other day," he said. He sauntered back to his pickup, giving me one more sidelong glance before he drove off. I took note of his warning about rattlers and watched where I stepped the rest of the time I was there.

∽

When I finished my surveys after three summers, I had lots of data on birds and their habitats for sites spanning a wide range of grasslands. I sat down to search through the data for interesting patterns. Nothing emerged. The sites varied in which bird species were present, the number and diversity of species, and many of the measurements of habitat structure. Yet, aside from the obvious conclusion that tall-grass prairies had more vegetation of higher stature than shortgrass prairies, there was nothing much to see. The differences among the sites seemed haphazard—not at all what I expected from prevailing theory (or from Martin Cody's earlier results).

I was disappointed. I felt that I had failed to advance our understanding of how grassland bird communities are organized. My observations simply did not align with the predictions of the competition paradigm. Perhaps I was asking the wrong questions. At this point, however, I was wondering whether this pathway had come to an end. Would I be able to obtain additional research support from NSF after failing to answer any questions that might interest my scientific colleagues? A conceptual framework to explain my observations would emerge only several years later.

In the meantime, our family had grown. David was born in 1969. I began to wonder whether my roaming around the grasslands in the summer was fair to my family. Unless they are unattached, most ecologists who do fieldwork face this dilemma. Some take their family along into the field, but this generally works only if the fieldwork is at a single location. Mine was spread across several states and I spent only a few days at any one place. Doing this sort of ecology requires that you have an understanding and supportive partner. Martha knew what the pathway I had chosen entailed and did her part to make it possible. But I did try to minimize the time that I was gone. Fortunately, the birds had a relatively short breeding season, so I was in the field only a few weeks at a time. Still, it wasn't easy.

7

EXTENDING THE PATHWAY
The International
Biological Program

Keeping a research program going requires continued funding, and continuing funding requires that, sooner or later, you must produce results that excite or intrigue the peers who will review your proposals. Coming up with inconclusive findings is not conducive to sticking to a pathway. My failure to add support to the competition paradigm, or to suggest how it could be modified or extended, placed my work well outside the domain of normal science.

My journey along the pathway of grassland birds could easily have ended at this point, but for the International Biological Program. The International Biological Program (IBP) was a broad, collaborative program that aimed to understand how productivity of the world's ecosystems related to human welfare. Although some IBP activities focused on how people adapted to environmental change and how biological resources could be managed for people, the primary emphasis was ecological. It was an audacious attempt to mount a comprehensive and detailed program to understand the mechanisms that controlled ecosystem processes and productivity. Thousands of investigators in countries throughout the world were involved (including, at the height of the Cold War, participants from the Soviet Bloc as well as the West). Planning for the program began in 1964 and the program continued for a decade.

Participation of the United States in IBP was organized around six broadly defined ecosystem types or biomes: tundra, deciduous forests, desert, coniferous forests, tropical rainforests, and grasslands. Although the biome programs differed in organization and focus, all aimed to understand the physical and biological components of ecosystems and how they contributed to ecosystem productivity and functioning. This objective would be addressed by conducting simultaneous studies of many ecosystem components at primary study sites representative of each biome. The grassland biome site was Pawnee, in the shortgrass prairie of northeastern Colorado.

Field studies in the Grassland Biome program were just getting underway in 1968. This gave me an opportunity to add the shortgrass prairie at Pawnee to my network of grassland bird-community sites. Although I was not studying ecosystem dynamics and energy flows (and in fact had little interest in them), the Biome staff was happy for me to work at Pawnee provided I would share my data with them. So I spent several days (and nights, camping) in the grasslands at Pawnee, mapping bird territories and measuring habitat.

The Pawnee site was part of the much larger Pawnee National Grassland, which extended for miles in all directions. It was a vista unlike that in any of my other study areas—grassland and open skies as far as one could see. The evening sunsets were spectacular, and the morning flight songs of longspurs were magical. I would have stayed there longer, but I had other sites to visit. I added the Pawnee data to my community comparisons. Perhaps I could come back someday.

"Someday" would soon arrive. That fall, I got a call from George Van Dyne, who was leading the IBP Grassland Biome program.

"I saw what you did at Pawnee last summer," he said. "We need a bird person to fill a gap in our data sets. How would you like to join our team?"

He went on to explain that the overall goal of the program was to build a huge computer model of an entire grassland ecosystem, using the approach of systems ecology. The model would have four components: the abiotic factors (soil, climate, water); producers (plants, especially grasses); decomposers (bacteria and fungi); and consumers (animals, including birds). They were adding a series of satellite sites in grasslands throughout the Great Plains and the western United States to complement and provide context for the more intensive studies at Pawnee. I would survey the birds at these sites to provide data for the consumer component of the model, but I could also use the data in my bird-community comparisons. I would be able to conduct more intensive studies at Pawnee. "And," he said, "we'll provide three years of funding, including some support for students."

Take a moment here to consider how things came together to determine my pathway at this point. I wanted to continue studying the ecology of grassland birds, but my current grant support from NSF was ending and I would need additional funding. My efforts so far, however, had failed to produce results that would impress my scientific peers. Then an opportunity appeared, with propitious timing. The IBP Grasslands Biome program was just getting started. George Van Dyne was looking for a bird person, and I had just surveyed the birds at the program's primary study site. George's invitation would provide the funding I needed. I would also be able to join a large research team involved in the first "big science" program in ecology, which would help me become part of the broader scientific culture of ecology. The emphasis on ecosystems might create an opportunity to break away from the grip of the competition paradigm. Given the objectives of IBP and the Grassland Biome's interest in grazing practices, my work might even have some broader societal relevance. For all of these factors to come together at the same time was truly serendipitous.

It was an opportunity I couldn't pass up. Now I would be able to spend several more years studying grassland bird communities. I could continue to roam the West, adding to the data and observations from the sites I'd visited under NSF support. Perhaps then I could answer some of the questions I had been asking about grassland bird communities: What habitat features relate to variations in bird species diversity or abundance? Are there consistent patterns of organization of the bird communities? How (and why) do bird communities vary across the spectrum of grassland conditions? Are there patterns of niche differences or overlap among species that are consistent with the predictions of competition theory?

The most straightforward way to address such questions is by looking for simple correlations: does some aspect of the bird community or the abundance of a species vary (positively or negatively) with variations in some habitat variable or the abundance of another similar species? With a large data set assembled from sites spanning a wide range of grassland conditions, I should surely be able to detect patterns of correlation. Or so I thought.

EPISODE: BIRDS ON THE IBP SITES

I began to survey birds on IBP grassland sites in the summer of 1970, joined now by graduate student John Rotenberry.[1] We followed much the same protocol that I had used in the NSF studies: arrive at a site, set up a grid of stakes in a plot, map bird territories, measure habitat features, and then move on to the next site. The Grassland Biome research design included consideration of the effects of grazing by cattle or native herbivores (e.g., bison), so at most sites we surveyed two plots that differed in grazing intensity or seasonality. This increased the number of plots we could analyze, but it also enabled us to compare plots with different habitat structure at a single site. At some sites we surveyed replicate plots that received the same grazing treatment to assess local variations in bird-habitat relationships.

Birds were only a small part of what was being studied at the IBP sites. Other teams of investigators were gathering data on small mammals, insects, vegetation, fungi, and physical properties such as soil chemistry, water balance, and temperature regimes. Consequently, local people were accustomed to seeing groups of researchers swarming through the grasslands, doing strange things. Still, our activities produced unusual puzzlement.

One of the shortgrass prairie sites, Pantex, was on the grounds of the Pantex Plant in the Texas Panhandle. The Pantex Plant was (and still is) a U.S. government facility charged with the production, dismantlement, and maintenance of nuclear weapons. Obviously, security was a major concern. After obtaining security clearances, John and I set up our study plots and began the bird surveys. In addition to our normal activity of pacing off grid locations and driving wooden stakes into the ground, John was making behavioral observations as part of his graduate research. Drawing on my approach to observing behavior in the Wisconsin marsh, John set up a ladder, climbed to the top, and sat there for a few hours using binoculars to record behavior while I chased birds to map their territories.

We had noticed earlier that we were being watched from a security vehicle parked some distance away, but we shrugged it off. After some time, however, the truck rapidly drove up. Two security officers got out. One came over saying, "I can't stand it anymore. We're supposed to leave you alone and just make sure you don't do anything suspicious. But what the hell are you doing? What's up with the guy with binoculars on top of the ladder?"

I explained what we were up to, how it fit into the broader research program of the Grassland Biome and assured him that we were harmless scientists. It turned out that he was a birdwatcher, so he understood at least some of what we were doing.

"OK," he said, "y'all be careful. And, oh," he added, "you might want to avoid the bunkers. They're full of rattlesnakes." The bunkers were large, dirt-covered sheds that had been used to store munitions but were now abandoned.

"We will," we quickly assured him. And we did.

He kept watch over us for the duration of our work at Pantex and waved to us when we left.

John and I surveyed birds and their habitats at the grassland sites for three summers. This would surely be time enough to discern patterns and answer our initial questions. Sure enough, there were differences in the bird communities among the sites.[2] Tallgrass prairies, with their knee-high grasses and forbs, supported a different complex of species than the closely cropped shortgrass prairies. Even within a site, areas with different grazing regimes contained different species. At Pawnee, for example, lark buntings and Brewer's sparrows occurred on a plot that was heavily grazed in winter, while a plot that had heavy grazing in summer (and thus had shorter vegetation and less litter) supported only McCown's longspurs. Although the distribution and abundance of several species were correlated with one or another of the habitat features we measured, no clear patterns of niche complementarity emerged, nor were community measures such as bird species diversity consistently related to habitat differences among the sites. My hope that adding the IBP grassland sites to my earlier comparisons would reveal clear patterns of bird community organization was in vain.

Faced with this discouraging conclusion, I had to ask why. Competition-based community theory assumed that available resources such as habitat and food were fully utilized and that competition reduced niche overlap among species to a permissible level. The abundance of a species, differences among coexisting species, and the composition of local communities should then be stable (i.e., in equilibrium) and suitable habitat should be fully occupied. Variations in species composition and abundances among locations, or over time, should closely follow variations in the availability of habitat and resources.

Because we surveyed most IBP sites over two to three years, I could assess the effects of both temporal and spatial variation. If I considered just the number or overall diversity of bird species present in a plot, the bird communities were indeed relatively stable from year to year, as Martin Cody had found in his more limited comparisons. But a closer look at the data from my repeated surveys showed that the abundance of most species in a plot varied, sometimes dramatically. One year a plot might contain territories of only one or two singing males and the next year it might be packed with territories. For example, dickcissels were scarce in one tallgrass plot in 1971 but ten times that number were present in 1972, even though the habitat had changed little and other factors such as weather were similar. In some cases, the abundance of a species even changed in opposite directions between years in nearby plots at the same site. Densities of grasshopper sparrows at the tallgrass site, for example, decreased by a third in one plot from 1971 to 1972 but were half again greater in a nearby plot. Overall, the picture that emerged was one of considerable variation in time and space, eroding any close relationships with habitat measures expected from theory. At the scale of our survey plots, available habitat did not appear to be fully occupied.[3]

The focus throughout these studies had been on habitat. I had measured habitat variables in detail, expecting patterns of niche differences among species to emerge. But differences among species might be expressed in a different way. Perhaps the species, instead, coexisted in the same habitats by differing in the food they ate. Food had always been considered a critical resource over which species might compete, but ecologists often assumed that differences in habitat occupancy and use would lead individuals to encounter different prey, and thus reduce diet overlap. Robert MacArthur had made this assumption in his studies of warblers in eastern forests. Even if species occupied the same habitat, however, they might still select different prey, especially if the species differed in morphological features related to feeding.

In fact, differences in bill size or body size among bird species had often been used as surrogate measures of niche partitioning. In the late 1950s, G. Evelyn Hutchinson had calculated the ratios of bill sizes for several sets of coexisting, ecologically similar bird species.[4] Adjacent species on a bill-size gradient had a fairly consistent ratio of 1.3; one species would have a bill 30 percent larger than the smallest species, the bill size of the next largest species on the size gradient would be another 30 percent larger, and so on. Hutchinson concluded that the value of 1.3 might indicate the difference (30 percent) necessary to permit the coexistence of such species. A bill-size ratio of 1.3 came to be regarded as somewhat of a constant in the predictions of competition theory.

Following this suggestion, my attention turned to morphological and dietary differences among species. The birds occurring on my study plots exhibited a wide range of bill shapes and body sizes, from long-billed (22 mm) meadowlarks weighing 112 grams to 17-gram grasshopper sparrows with short (8 mm) conical bills. Bill-size ratios for progressively larger species in these communities varied from 1.03 to 3.19, far from the regularity expected in theory. In fact, there was a conspicuous gap in bill sizes between a set of small birds and larger species such as meadowlarks, resulting in large bill-size ratios (2.50 on average).[5]

The use of size ratios to infer competition and niche partitioning assumes that morphology (e.g., bill size) should be closely related to ecology (e.g., prey size). Part of my responsibility for the Grassland Biome work involved collecting birds for diet analysis. In the next chapter I'll explain why I was doing this, but at this point suffice it to say that it provided information that could be used to assess similarities and differences in the diets and prey choices of birds in my study areas. For two or three years, John Rotenberry and I sampled the diets of the major breeding bird species at IBP sites in tallgrass, mixed-grass, and shortgrass prairies and in western shrubsteppe. We preserved the birds' stomachs and gizzards and later, hunched over microscopes in the lab, we meticulously examined their contents, identifying fragments of insects and seeds and estimating their size and biomass to catalog what the birds had been eating.

Summed up over multiple individuals, these analyses can yield an approximation of the diet composition of a species at a site. But it's only an approximation. Although we collected individuals at the peak of the local breeding season when competition might be most intense, the samples provided only a snapshot of a species' diet. Still, if there were clear and consistent differences in the diets of coexisting species, they should show up.

Although the stomach contents of the species at a site did differ, there was little pattern to the differences.[6] There was substantial variation among individuals of a species at a given site, among years at a given site, or among sites in a given year. Diet overlap among the species at a site was inconsistent and unrelated to differences in the birds' body or bill sizes. For example, small grasshopper sparrows had relatively high diet overlap with larger western meadowlarks in all three years of sampling at the mixed-grass site, while their overlap with eastern meadowlarks at the tallgrass site was high in some years but low in others. Overall, our results showed little agreement with the predictions of competition theory. Rather, the variations among species, sites, and years seemed to reflect opportunistic foraging. What we found in an individual's stomach was what it had encountered in its most recent foraging at a site, and most of the species were encountering much the same prey at the same time.

My participation in the IBP gave me the opportunity to consider multiple features of grassland bird communities at a broad array of sites over several years. I had plenty of data. Try as I might, however, I couldn't seem to mesh my observations with the expectations of resource limitation and the predictions of the competition paradigm. There was no evidence of community equilibrium. Instead, I was seeing variation and overlap among species everywhere I looked. I was trying to fit the square peg of my observations into the round hole of competition theory. Perhaps it was time for some fresh thinking.

8

SCRAMBLING FOR AN EXPLANATION

Climatic Instability and Ecological Crunches

One of the features of a paradigmatic science is the way in which the paradigm, once established, dictates how you will interpret the results of your studies. This is what normal science is all about. If your observations fail to match the predictions that are guiding research in a discipline, what do you do? Such anomalous results are not really challenges to the paradigm if they can be explained away. So you look for plausible ways to reconcile them with predictions of the paradigm.

The simplest explanation is that your observations are biased or faulty. Of course, no scientist likes to admit that their data might be in error. John Rotenberry and I had paid close attention to how we designed sampling and had taken pains to ensure that our methods were consistently applied. Unless the consequences of competition were very subtle, we should have detected them. We considered this explanation unlikely.

Alternatively, perhaps the theory was simply wrong and the predictions invalid. But competition theory had become a well-established paradigm in ecology. It was unlikely that so many ecologists would have accepted the theory had they not found patterns that were consistent with the predictions. Instead, we kept coming up with the anomalous observations that Kuhn had talked about.

Perhaps the problem was with the underlying assumptions. All theories make assumptions—if these hold, then the predictions logically follow. Perhaps the bird communities we had studied violated the assumptions of competition theory. Somewhat reluctantly, I concluded that it was time to reassess the paradigm and its underlying assumptions.

Competition theory makes several important assumptions. One, of course, is that the species considered are ecologically similar and are potential competitors. One wouldn't expect competition between species having little potential overlap like, say, a hawk, a swan, and a sparrow. However, all of the species we studied occupied similar grassland habitats, had overlapping territories, foraged in similar manners, and fed on roughly the same things. Such species would be expected to compete if critical resources were consistently in limited supply relative to demand. Under such conditions populations would grow to reach an equilibrium level determined by resource availability. Resource limitation would then exert a continuous and intense pressure to reduce niche overlap and thereby circumvent direct competition. The result would be the sort of niche partitioning and community equilibrium predicted by theory.

But what if resources were not limiting, but were instead superabundant relative to demands? In this case, species that overlapped broadly in their opportunistic use of resources would pay no penalties; there would be plenty for all. The patterns expected from competition theory would not emerge. If resources were consistently superabundant, however, populations should expand to capitalize on the opportunity and additional species might be able to join the community. The community would then reach the point where resources did become limiting and the predictions of theory would hold.

I reasoned that conditions in which resources were not limiting, competition not ongoing, and niche partitioning not evident might be most likely to occur in variable environments, where resource levels varied from feast to famine.[1] Periods of resource abundance would be punctuated by times of resource scarcity—"ecological crunches." During ecological crunches the number of species in the community and their population levels might be restricted by resource limitation and competition. The expected patterns of niche partitioning would emerge. At other times, however, competition would be relaxed and niche overlap between species, and variation among individuals within species, would increase. The predictions of competition theory would be expected to hold during ecological crunches, but not at other times. Populations with a relatively low capacity to respond to the variations in resource abundance would be unable to grow quickly to take advantage of abundant resources as conditions improve following an ecological crunch, so resources might often be superabundant relative to demand. Birds would fit this model better than, say, insects.

Thinking that grasslands might meet these conditions, I looked for a suitable measure of environmental variation that might be related to resource abundance. In many ecosystems, but perhaps especially in grasslands, primary (plant) production is closely tied to precipitation. Variations in primary production, in turn, affect habitat structure and the biomass of insects and seeds over which breeding birds might compete. Extreme periods (e.g., droughts) might create ecological crunches.

To explore this possibility, I analyzed long-term records from weather stations in tallgrass, mixed-grass, and shortgrass prairies.[2] These grassland types represent a gradient of decreasing annual rainfall: the tallgrass stations averaged half again more rainfall than the mixed-grass stations, which in turn averaged one-third more than the shortgrass stations. Plant production varied accordingly. The important measure, however, is the variability in rainfall from year to year. Annual variation was significantly greater in shortgrass than in mixed-grass stations, which in turn were significantly more variable than tallgrass stations. In shortgrass areas, roughly one out of every twelve years could be expected to be extremely dry or wet (i.e., to deviate from the long-term average by at least half that average). In mixed-grass prairies, on the other hand, extreme years were only 60 percent as frequent as in shortgrass, and such conditions occurred on average only once every forty-two years in tallgrass prairies. Ecological crunches may be frequent in shortgrass prairies and infrequent in tallgrass prairies.

As far as they went, the results of our surveys of bird communities over the range of grassland types were consistent with the ecological-crunch hypothesis. Although the number of species and species composition of local communities were relatively stable, abundances varied considerably, resource use appeared to be opportunistic, and species overlapped in habitats and diets. But these surveys were conducted for only a few years. To really test the hypothesis—to determine whether there is a relationship between climatic variability, the frequency of ecological crunches, and the composition, stability, and niche partitioning of grassland bird communities—would require long-term observations, and the tallgrass surveys might need to continue for a very long time to encounter a crunch. Such data are not available. However, our IBP surveys over three years at the shortgrass site at Pantex did include a severe drought and an unusually wet year. Species' abundances varied accordingly. Horned larks were abundant during the drought but nearly absent in the wet year, while grasshopper sparrows showed the reverse pattern. Ecological overlap among the species, however, was not noticeably less in the drought year. Perhaps the drought was not severe enough, or long enough, to represent an ecological crunch.

~~~

The IBP had given me the opportunity to explore how the composition of grassland bird communities and their habitat, morphological, and dietary dimensions varied in space and time. Our results cast some doubt on the ubiquity of the competition paradigm. While competition theory might not itself be wrong, it simply might not apply in situations where its assumptions were not met. The alternative ecological-crunch hypothesis suggested that conditions that limit habitat, food, or other resources might occur infrequently in variable environments. Resources might often be superabundant relative to demands, competition would be relaxed, and the patterns expected from theory would not be evident.

I was excited to develop this line of thinking more fully. But in the meantime, my involvement in IBP had spawned another line of research. Another pathway beckoned.

9

DETOURING TO ANOTHER PATHWAY

Modeling Bird Bioenergetics

S upport from the IBP Grassland Biome program allowed me to continue searching for evidence of competition in grassland bird communities. The questions that guided this research—about habitat relationships, niche overlap, and how bird communities varied in space and time—were the questions I had been asking for years. They were *my* questions.

But they were not the questions that interested the Grassland Biome program. The program was supporting my research to address their questions, not mine. And their questions were about the productivity of grasslands: how energy and nutrients were transferred among the components of grassland ecosystems; how these patterns differed among the major types of grasslands in North America; and how this all related to human uses of grasslands, particularly grazing. These were entirely different questions. They would be addressed by building on the growing application of systems analysis methods in ecology to construct computer models of the dynamics of grassland ecosystems.

I was supported by the Grassland Biome to provide the data and analyses to quantify the role of birds as consumers in grassland ecosystems. This was the string attached to the funding. When IBP support began in 1969, I began to ask questions that were embedded in the broader framework of the Biome program. What were the patterns of production and energy flow through bird populations? How did these differ between tallgrass and shortgrass prairies, or under different grazing regimes? What role did birds play in the bioenergetics of grassland ecosystems? The source of funding rather than my own interests dictated the questions I would ask.

These questions opened the door to another pathway. Because it would not be possible to directly measure the energy demands of active, free-living consumers such as birds, estimates would need to be derived by modeling. This pathway, then, would rely on modeling instead of the observations and theory that had guided the bird-community pathway.

But there was an immediate problem. Although I could envision the general features of what a model of bird bioenergetics might look like, I had little facility with computers or programming. Creating a working computer model would be well beyond my ability. Things might have stopped there, but events intervened. In 1970, George Innis joined the Grassland Biome program to direct the modeling efforts. George and I hit it off and quickly began to collaborate. I would provide the information about birds necessary to conceptualize a model and the data to populate it, while George would write the computer code and operate the model to make it work. So George and I set off on the modeling pathway.

EPISODE: DEVELOPING A BIRD BIOENERGETICS MODEL

Ecosystems are tremendously complex affairs, with a great many constituents that interact in a bewildering array of ways. To deal with this complexity, the Grassland Biome program used computer simulation models to partition a system into its constituents, their interrelationships, and the dynamics that the interactions produce.

Such models are often visualized as "box-and-arrow" diagrams, in which the boxes are the system constituents and the arrows specify their interactions. The interactions, in turn, generate changes in the constituents over time. All of this can be specified by mathematical functions, using energy flow and production as a common currency. George and I adopted this approach in dealing with birds in grassland ecosystems.

Birds are linked to the dynamics of an ecosystem by their individual energy demands, the foods they consume to meet those demands, and how many individuals there are—their population dynamics. The model that George and I developed (creatively labeled BIRD) included two component models, one dealing with the energy demands of individuals, the other with population dynamics and demography. By combining the outputs of these submodels with information on food habits, the demands that birds placed on their prey could be estimated in energetic terms. This, in turn, would indicate the role birds might play in the energy dynamics of grassland ecosystems and would provide the information needed to include birds in the comprehensive grassland ecosystem model.

The details of the BIRD model would be mind-numbing to anyone other than a modeling aficionado.[1] In essence, we calculated daily energy demands per individual for each species at a site using information on ambient temperature, body weight, and metabolic equations from the literature, adjusted for the additional energy costs of various activities.[2] The demographic submodel used my estimates of the population density of a species at a site, adjusted to reflect seasonal movements, mortality, and reproduction. Combining the outputs of these two submodels gave us an estimate of daily energy demand for the population of each species at a site. Using the information on what birds were eating and the energy content of different prey items, we could then translate the population energy demands into estimates of the total energy flow to birds from several prey categories. This was the information needed to indicate the role of birds in the energy dynamics of the grassland ecosystem.

What did all this modeling tell us? Our primary aim was to produce estimates of energy flow to birds at the IBP grassland sites.

BIRD did this. The patterns were pretty much what one would expect—energy flow to birds was greater at sites with more birds, especially if the birds were larger (and therefore required more energy). There were no consistent relationships to grazing intensity, but there were some unexpected patterns. For example, the total April-August energy flow to breeding birds at a tallgrass site in Oklahoma (Osage) was the same as that at a shortgrass site in Texas (Pantex), despite differences in the bird species present and the growth form and productivity of vegetation. The birds at the two sites, however, obtained their energy from different sources: seeds contributed 51 percent of the energy demand at Pantex but only 12 percent at Osage. However, over all of the grassland sites, the total breeding-season energy flow to birds was small.

According to these modeling results then, it seemed unlikely that birds would have any significant effects on ecosystem dynamics in these grasslands through their role in energy flows. Rather, it appeared that the flush of ecosystem production in summer would normally exceed the capacity of birds to utilize food resources, at least under the conditions that we modeled. George and I concluded that birds may be "frills in the ecosystem," living and reproducing off the seasonal bounty without really influencing the broader picture of ecosystem energy dynamics. These model results came from an entirely different direction than my community analyses, but they supported the same conclusion: energetically, birds breeding in these grasslands were apparently not limited by their food resources.

The BIRD model was designed to estimate the bioenergetics of birds in grassland ecosystems. Its structure was general enough, however, to allow us to apply it to explore energy-flow patterns for birds in other situations. Over the next few years, I collaborated with several scientists to generate values for the many input variables required by BIRD. For example, Ron Nussbaum provided information from the IBP Coniferous Forest Biome program on bird populations breeding at six forest sites in Oregon that varied in elevation, temperature, and moisture.[3] You don't need to be an ecologist or ornithologist to see that there are more breeding birds in a coniferous forest than in a grassland. The model estimates suggested that energy flow to breeding birds in the coniferous stands was perhaps eight times greater than that in our grassland sites. Lacking information on bird diets or the abundance of prey available to birds in the forests, however, Ron and I could not address the broader question of whether the bird communities were energetically close to resource limitation or whether the ecological-crunch hypothesis was likely to apply.

In a second application of BIRD, Mike Scott and I modeled the energy flow through populations of four seabird species over the entire Oregon coast for an entire year.[4] Brandt's cormorants, common murres, and Leach's storm-petrels breed along the coast and winter at varying distances from the shore; sooty shearwaters breed in the Southern Hemisphere and occur on the Oregon coast primarily as spring and fall migrants, when they vastly outnumber the other species. Consequently, they overwhelmingly dominated energy flows, consuming seven times more energy per day than any of the other species.

Mike had studied seabirds for his PhD and could supply information on the diet composition of the seabird species and the caloric content of prey. Using this information, we could convert energy flow to estimates of total prey consumption. Over the year, we estimated that the four species would consume some 62,500 metric tons of prey, of which 28,000 metric tons were anchovies. Most of the anchovies (85 percent) were consumed by shearwaters during their relatively brief stay in coastal waters.[5] Overall, Mike and I estimated that the four seabird species might consume as much as 22 percent of the pelagic fish production in the Oregon coastal zone.

The numbers, of course, are model estimates, subject to the many assumptions of the modeling. However, they do help to define the ballpark within which birds may operate as consumers in these ecosystems. And in this case, the numbers supported the suggestion that seabirds may play a significant role in the energy dynamics of coastal marine ecosystems, much greater than that of breeding birds in grasslands or northwestern coniferous forests.

In one other application, Mel Dyer and I used BIRD to estimate food consumption by red-winged blackbirds in a large nighttime roosting aggregation (225,000 birds) on the shore of Lake Erie.[6] During the day the birds dispersed from the roost to feed on corn and other grains in surrounding areas, avoiding feeding in areas close to the roost where control efforts by farmers were greatest. Mel and I calculated the estimated energy demands of blackbirds at increasing distances from the roost and then converted the energy demands into estimates of overall consumption of insects, corn, detritus from wheat and oat fields and weeds using information on the birds' diets and the caloric values of each food type. Insects and wheat-oat detritus dominated the diets at the beginning of the roosting period in late July, but as the corn crop matured in late August the birds shifted to corn. This is when their potential agricultural impact was greatest. Over the entire roosting period (late July until the end of September), we estimated that corn would comprise 30 percent of the total biomass of food consumed by the blackbirds, the majority in an area 12 to 35 km from the roost.

By converting the estimated corn consumption into bushels of corn, we could then use information on crop acreages and yields for the area that we modeled to calculate the amount of corn available to the birds at various distances from the roost in relation to what they consumed—their potential impact. We estimated that consumption of corn by the blackbirds represented less than 1 percent of agricultural yield at distances greater than 40 km from the roost but as much as 16 percent close to the roost. This is enough to have an economic impact. Recognizing this, some farmers with fields close to blackbird roosts switched to crops that are less palatable to the birds. In some parts of the world, explosives are used to destroy roosts (and birds) that are wreaking havoc on grain crops.

EPISODE: POLAND AND GRANIVOROUS BIRDS

The application of the BIRD model to blackbirds and corn consumption provided us with an opportunity to join with another IBP program that was broadly international in scope. The Working Group on Granivorous Birds brought together more than one hundred scientists from the United States, Europe, Africa, India, and elsewhere to synthesize information on the population dynamics, food habits, and potential impacts of granivorous (grain-eating) birds on agricultural ecosystems. Most of the studies dealt with two species of sparrow (*Passer*), the ubiquitous house sparrow (*P. domesticus*) and the European tree sparrow (*P. montanus*). The information assembled by the working group enabled Mel and me to use BIRD to estimate the energy demands of populations of these two species under a broad range of conditions.

Polish scientists, led by Jan Pinowski of the Institute of Ecology, Polish Academy of Sciences, coordinated the activities of the working group. As is often the case with large international working groups, there were many meetings to gather and analyze data. Mel and I each hosted a meeting at our home institution, but most were held in Poland. At that time, in the late 1960s and early 1970s, the Cold War was easing, détente was growing, and Polish and Eastern European scientists were as eager to collaborate as were their Western counterparts. At a meeting in Dziekanów Lesny, Poland in 1973, for example, I shared a room with Victor Dol'nik, a scientist with the Soviet Academy of Sciences in Leningrad. Although he knew little English and I only a few words of Russian, we managed to spend hours sharing ideas about birds and energetics. In the evenings, fortified by Polish vodka, we found that our linguistic fluency somehow improved, and we often talked politics.

Vodka also figured prominently in the evening gatherings of the entire group. Polish vodka is particularly potent, and we Americans had a relatively low tolerance threshold. I well remember how Kazimierz Petrusewicz, the dean of Polish animal ecologists, would stroll through the group holding a shot glass, looking for Americans. Spotting one, he'd raise his glass in a toast, a challenge one couldn't refuse. He'd down his shot, refill, and resume his search for victims while we struggled to remain sober. We soon learned to avoid direct eye contact.

But mostly we went about our business. Through our collaborations with scientists in the working group, Mel and I pulled together data to conduct model simulations for twenty-four *Passer* populations in thirteen European countries.[7] Because our primary aim was to estimate how energy demands might vary between the species and among locations, we standardized comparisons by considering energy demands of a population of one hundred adults km^{-2} at each location. On this basis, annual energy demands of *P. domesticus* varied from 1.4 kcal m^{-2} year^{-1} in a Wisconsin study area to 2.2 kcal m^{-2} year^{-1} at sites in Poland and the German Democratic Republic (GDR). The annual energy demands of *P. montanus* were generally similar but also varied among locations, perhaps because of differences in environmental conditions and the length of the breeding season.

As in our other exercises using the BIRD model, we coupled the energy demand estimates with information on diet composition and the caloric value of food types to estimate total annual food consumption. A standardized population of *P. domesticus* might consume 287 kg dry weight km^{-2} year^{-1} at a site in Missouri but nearly 400 kg dry weight km^{-2} year^{-1} at sites in the GDR and Poland. Estimated annual food consumption by *P. montanus* varied from 250 kg dry weight km^{-2} year^{-1} in northern Poland to 322 kg dry weight km^{-2} year^{-1} at a site in Romania. Cereal grains dominated food consumption by *P. domesticus* in all areas while consumption patterns of *P. montanus* were more variable, perhaps reflecting local variations in food availability. These results align with the observation that *P. domesticus* is more granivorous than *P. montanus*, and thus more likely to have impacts on grain agriculture.

Over several years, BIRD demonstrated the usefulness of a modeling approach to estimate the energy demands of different species in different settings in a consistent fashion. Combined with information on food habits, we could estimate food consumption rates and evaluate the role of birds in ecosystem energetics or their potential impact on resources important to people, such as corn or marine fish.

Modeling also revealed troublesome gaps in our knowledge. The energy calculations in the model, for example, were based on metabolic studies of birds in the laboratory. Converting these values to estimates of energy flow for different species in the wild required assumptions about the behavior of the birds, energy costs of various activities, digestion rates, diet composition, and an array of other variables. Tests could be conducted to determine the sensitivity of model estimates to such assumptions and prioritize further studies to fill in the gaps. As is often the case, however, the variables to which the model was most sensitive were the very ones most difficult to measure. And because BIRD was used to estimate values that could not be obtained in nature, it was difficult to validate the model—to ensure that the estimates corresponded with what was actually going on in the real world.

I had other concerns about the overall thrust of our modeling. By collapsing the important dynamics of ecosystems into calculations of energy flow and productivity, many of the features of natural history and behavior that encompass the vibrancy and beauty of ecosystems were left behind. As in any computer simulation model, the boxes and arrows of BIRD simplified reality by compartmentalizing components of the system and streamlining their interactions. This was necessary to make the model work. But it resulted in an incomplete rendering of birds in grasslands—of nature. It's not just that an ecosystem model cannot capture the magic of a morning flight song of a longspur or the quiet radiance of an evening sunset over the prairie. After all, these are ecosystem properties that cannot be expressed as model functions in quantitative terms. But other more clearly ecological features were reduced to simple calculations. Thus, although George and I included population dynamics in the BIRD model, they were there only for accounting purposes. They provided a tally of individuals that could be translated into energy demands by different age classes; there were no functions that described what might regulate populations or that captured predator-prey or competitive interactions. There were no functions relating habitat conditions to the behavioral responses of species, or to express how population densities might vary among years in different ways for different species in different places. In short, the sorts of things that had so interested me in my earlier studies of grassland birds were missing. Behavioral ecology was largely absent.

This should not be taken as a criticism of the Biome program. The program had different objectives, and the comprehensive grassland model that it created helped place ecosystem studies on a firm scientific foundation that set the stage for their subsequent growth as part of the mainstream of ecology. It was also a source of many productive collaborations.

Collaborations can play an important role as you become an ecologist. Ecological systems are complex, with multiple interconnections among disparate parts. Understanding them often requires complementary expertise. Consequently, you shouldn't fall prey to the notion that you can do everything yourself. By contributing expertise that you simply don't have, collaborators can help you answer questions you would otherwise be unable to address. But collaborations involve people and, inevitably, personalities. When personalities mesh, collaboration can be built on mutual respect. You enjoy the collaboration; it is a delight. If personalities clash, as can happen when egos dominate, the collaboration can become a chore. Although such collaborations can be productive, you should ask whether you're having fun and if not avoid them if you can.

In my case, the collaborations were both fruitful and fun. But it was hard for me to become enthused about energy flow and productivity. Still, I could have continued on the modeling pathway, applying BIRD to estimate energy demands or food consumption by birds in many other situations. It was a productive pathway, one that generated many publications and opportunities to collaborate with colleagues from a variety of other countries. But the modeling was simply describing how much energy flowed to bird populations from various sources under several restrictive assumptions. This was not addressing the questions that interested me. There were no hypotheses to test, no central concepts, no overarching paradigm. It was desktop research. It didn't generate that deep, passionate interest that keeps one going on a pathway.

There's a broader lesson here. As the array of technologies and tools in ecology continues to expand, it's easy to be enticed into regarding a tool as the foundation of your research. Your questions shift from asking how organisms relate to the environment and one another (i.e., ecology) to asking how you can use the tool in a variety of settings—your questions are tool-driven rather than nature-driven. This can be very productive, to be sure, but it can easily come to dominate your research. That's fine if your real interest is in tools and technology (e.g., in using models such as BIRD). But if your interest is in exploring how nature works, use the tools when they help you but put them aside when you don't need them.

At some point I realized that the modeling pathway was leading me away from where I wanted to be and what I really wanted to do: conducting field work in open spaces, exploring concepts, and pursuing the nonequilibrium ideas that came out of the bird-community studies. It was time to return to my questions about the organization of bird communities. But the practicalities of life were exerting their influence once again. I wanted to work some place closer to my family and our home in western Oregon. The expanses of sagebrush in the northern Great Basin had many similarities to the grasslands that I'd been studying: an open habitat, variable climate, and a simple bird community of only four or five breeding species on a study plot. I could continue down the bird-community pathway in the sagebrush shrubsteppe.

10

MOVING FROM GRASSLANDS TO THE ARID SHRUBSTEPPE

I mages of Oregon usually feature coastal panoramas and the magnificent coniferous forests of the Coast and Cascade mountain ranges. Yet much of eastern Oregon is a high desert, dominated by sagebrush shrubsteppe. I had paid little attention to these areas as we hurried through them on our move west, anxious to begin my new job at Oregon State. I thought I might initiate research in the forests. Once again however, people, location, and circumstances were to open up a new pathway.

It began with one of the graduate students that I inherited on my arrival at Oregon State. Ralph Moldenhauer was conducting his doctoral research on sage sparrows in the arid shrubsteppe of the northern Great Basin, where they were common breeders. He was asking whether the sparrows had any special physiological adaptations for living without access to free water. Since he was working with birds, it seemed logical that I would take over directing his studies. Fortunately, other faculty members were available to help with the physiological details of the laboratory part of his study; I could help place his results in the ecological context of where the birds live.[1]

Shortly after I arrived, Ralph suggested that we visit the area in central Oregon where he captured birds for his laboratory studies. We set off one fall weekend, driving over the crest of the Cascades, through the ponderosa pine forests on the eastern slope of the mountains, and then suddenly emerging into the sagebrush expanses of the Great Basin. I was immediately enthralled. Here were the wide-open spaces I loved. I no longer felt hemmed in by all those trees. The sage sparrows had long since left for the winter, but I could easily imagine continuing the sort of studies I had done before—mapping territories, measuring habitat features, finding nests, and testing competition theory.

Although this pathway was enticing, it remained untrodden for a while. I would need grant funding to work in the shrubsteppe, and the prospects for success in getting NSF support would be greater if I continued the grassland research, where I was already beginning to establish a track record. Thoughts of working in the Oregon shrubsteppe were put aside while I expanded the grassland studies to multiple sites (and experimented with tadpoles during the winter).

But the idea of conducting research in the shrubsteppe was not entirely abandoned. The pathway to continue studies of grassland bird communities as part of the IBP led back, in a way, to the shrubsteppe. The IBP Grassland Biome network included one site, the Arid Lands Ecology Reserve (ALE), that was dominated by bunchgrasses and sagebrush typical of much of the northern Great Basin.[2] ALE was part of the Hanford Atomic Energy Commission Reservation along the Columbia River in eastern Washington. Public access to the site had been restricted since the early 1940s, when the Hanford facility began producing plutonium for atomic weapons in support of the Manhattan Project. Although security was still tight when we worked there in the early 1970s, we were able to conduct our studies with few interruptions. In fact, rather than setting up a camp and sleeping under the stars, we were allowed to stay overnight in an underground bunker for a missile silo that had been decommissioned.

This led to another interaction with government security officers, somewhat reminiscent of our experience at Pantex. We were
staying in a large room that had been part of the missile control center, but now was largely empty. To wind down at night, the three of
us began tossing a Frisbee about. We'd been doing this for a while
one evening, when a voice boomed out.

"What the devil are you guys doing?"

We looked around. There was no one there. Then the voice
boomed out again.

"Sounds like you're tearing the place apart."

It was Hanford Security, who'd been listening in. I suppose we
might have expected that our activities would be monitored, but
this was disconcerting all the same.

"Oh, we were just tossing a Frisbee back and forth," we rather
lamely explained. "It's not against regulations, is it?"

After a pause, the voice replied, "I don't think the regulations
cover Frisbees. Just don't do it."

We stopped our evening recreation and were careful what we did
or said for the rest of our time in the bunker.

Although the bird species breeding on our study plots at ALE differed from the communities at the more typical grassland sites in
the IBP network, the general patterns were much the same. There
were three or four breeding bird species at ALE, three to five at the
grassland sites. Total breeding density at ALE was on the order of
200 to 300 individuals km^{-2}, as in the grasslands, although the total
biomass of birds in a plot was less (the shrubsteppe species were generally smaller). It seemed that the bird communities in the shrubsteppe might provide a good way to assess the generality of what
we were finding in the grasslands.

Field studies of bird communities as part of the IBP Grassland program ended in 1974. It was time to obtain funding to continue my
research and support a growing group of graduate students. The

shrubsteppe pathway beckoned again. Changing directions or locations in a field-oriented research program, however, is not something to be undertaken lightly. You must deal with the perception that you have abandoned a pathway before answering the important questions (e.g., What are the patterns of niche partitioning among coexisting species? Do the community patterns support the prevailing paradigm?). You must convince your peers (the normal scientists) that the new location or study system is a logical extension of your previous research. You must not come across as a gadfly, flitting from one place or system to another without tying up the loose ends. Apparently, I succeeded in addressing these concerns. My proposal to NSF, "Avian Responses to Environmental Heterogeneity in Arid Shrubsteppe Systems," was funded. So along with several students (including John Rotenberry), I turned my attention to birds in the shrubsteppe.

The sagebrush-dominated shrubsteppe of the northern Great Basin was attractive for several reasons. Ralph Moldenhauer had shown me the shrubsteppe at Cabin Lake, and it seemed an ideal place to base our long-term studies.[3] It was only a few hours from the Oregon State campus. The sagebrush-dominated shrub desert was vast, extending unbroken for many miles. Most of the land was federally owned and used only for light and widely scattered grazing. Although several homesteads had been established in the area early in the twentieth century, they had long since been abandoned. There would be ample opportunities to establish multiple study plots. It was just a short walk from the study plots to a U.S. Forest Service guard station with a water source, where we could set up a seasonal camp. And, as it turned out, there were no snakes there.

This last point merits a brief comment. In more than a decade that our group spent summers at Cabin Lake, no one ever saw a snake or even found a snakeskin. Yet there was an active and prolific rattlesnake den in a rocky outcrop only a few miles away. Nights in the Oregon high desert can be cold, even in mid-summer (we once had several inches of snow on July 4th!). The Cabin Lake site is in a depression that receives cold air drainage from Paulina Peak some fifteen miles to the north. Consequently, most nighttime temperatures are close to freezing; the average frost-free period is (or was then) four days—much too cold for ectothermic reptiles like snakes. Because snakes are often major predators on eggs and the young of nesting birds, their absence from Cabin Lake resulted in unusually high nesting success for the bird species breeding there. Yet, although we banded hundreds of fledglings over the years, we never found any returning young. They either all died or, more likely, settled somewhere else in the sagebrush expanses to breed. As far as we could tell, the lack of snake predation had no effect on local bird abundances or community organization.

But I did enjoy working in a snake-free environment.

Our studies at Cabin Lake began by asking some of the same questions we had initially asked in the grasslands. What were the characteristics of habitats occupied by the breeding bird species? What were the patterns of overlap among species in habitat occupancy or diet composition? How did diets relate to differences in morphology? How did these patterns vary over time or space? Was there clear evidence that competition structured the bird community? I was still assessing niche overlap among coexisting species but was no longer clinging so tenaciously to the competition paradigm to explain our findings.

The label "shrubsteppe" indicates the major way in which habitats in our study areas differed from those in the grassland sites that we had studied. The combination of shrubs with grasses and forbs gave the habitats greater structure and patchiness. In general, abundances of the characteristic shrubsteppe bird species—sage sparrows, Brewer's sparrows, sage thrashers, and black-throated sparrows—did not vary systematically among plots or sites in most measures of habitat structure.[4] However, the birds did seem to pay attention to which shrub species dominated a plot. Abundances of sage sparrows were (appropriately) greater on plots with more big sagebrush. Brewer's sparrow and sage thrasher abundances were unrelated to sagebrush coverage but decreased as coverage of two thorny shrubs, hopsage and bud sage, increased. Black-throated sparrows were more abundant on plots with more hopsage and greasewood. The strong associations of these bird species with shrub species rather than habitat structure hinted at a possible link to food resources associated with the shrubs—an inkling of a theme we were to pick up a few years later (see chapter 15).

The three years of our initial studies in the shrubsteppe included one of the driest years on record, followed by one of the wettest, so we could assess how (or whether) these variations affected the birds and their habitats.[5] The shrub species, which grow slowly, showed no detectable responses to the variations in precipitation. Although some structural features of the habitat, such as the height and coverage of forbs and grasses, changed substantially, the abundances of several bird species varied independently of any changes in habitat structure or other bird species. The changes in bird abundances seemed more consistent with a reshuffling of individuals on plots that were not fully occupied.

To broaden the scope of our studies in the shrubsteppe, in 1977 John Rotenberry and I established several supplemental "rapid survey" sites scattered throughout southern Oregon and northern Nevada. Rather than adhere to the niceties of randomized sampling, we selected the sites subjectively, picking places that "looked good"— places that were dominated by sagebrush; supported breeding sage sparrows; were immersed in a much broader expanse of shrubsteppe; were used only for grazing (if at all); and included a range of conditions characteristic of the northern Great Basin. At several sites we also surveyed a plot dominated by another shrub species (generally greasewood).

Our protocol was in some respects the antithesis of the detailed, intensive approach we took at Cabin Lake. We needed to visit the sites within a narrow window of time at the peak of the breeding season, so we couldn't spend much time at any one site. To choose the sites we drove down a highway until we saw a good-looking expanse, took a side road (or more often, a dirt track) several miles away from the highway, and stopped when we saw what we wanted. We then set up a linear transect through the sagebrush, driving wooden stakes into the ground at regular intervals (yes, we were still doing that). The transect was the framework for surveying bird populations and measuring habitat features. John and I surveyed birds in the morning and evening and measured habitat features during the day. We camped at the sites by finding a wide, level place in the track, where we set up a folding table and chairs, pulled up the cooler (stocked with beer), and tossed our sleeping bags on the ground. We stayed there for a day or two before heading off to the next site. No one ever bothered us, and we saw only one other person during our years of camping in the middle of the road.

Surveying the birds involved little more than walking the transect slowly, noting each individual of each species we saw or heard within 250 m on either side of the transect.[6] From this we could derive an estimate of the population density of each species, which we could then associate with the set of vegetation measures in the transect area. To characterize the vegetation, we recorded the same measures we'd been using elsewhere in the shrubsteppe.

Surveying the birds was a delight. The sagebrush expanses glowed in the morning or evening light, and there was no background noise to interfere with the singing birds. Measuring habitat was less exciting, although it did have its moments.

Which brings me to another snake story. To measure the coverage and height of the vegetation, we passed a thin metal rod vertically through the vegetation to the ground and recorded contacts of vegetation with the rod at height intervals. Generally this was routine, but once when I passed the rod through the vegetation I felt unexpected resistance, and then movement. I'd awakened a sleeping rattlesnake, which obligingly rattled. John said he didn't know I could move so quickly. We thought perhaps we'd best pick a different random location for that sampling point. To paraphrase Falstaff, "discretion is the better part of sampling."[7]

John and I continued the surveys for six years. The results matched those from our more intensive study sites: the abundances of the bird species varied but did so independently of changes in other bird species or habitats among the sites. Differences in habitat structure were not matched by the birds, although their associations with shrub species persisted.[8]

EPISODE: A LOOK BACK

By the time we finished the surveys in 1982, both John and I had moved on to other jobs and were following other pathways. But the shrubsteppe still beckoned. As 1997 approached, we realized that it would be twenty years since we had first visited the sites. Wouldn't it be a good idea, we rationalized, to revisit the sites and repeat the surveys to see what might have changed or whether the bird communities remained stable. And we could recharge our shrubsteppe batteries. So we returned to survey the sites again, using the same procedures and camping in the same places. We found the now-weathered stakes that had marked our transects, reset them, and used the same transects we had used before.

Apart from our now being twenty years older, not a lot had changed.[9] Over all the sites, coverages of shrubs, grasses, bare ground, and litter had increased slightly, but there were no significant changes in the coverages of shrub species. Among the birds, the abundance of loggerhead shrikes had decreased significantly (mirroring a range-wide decline), but abundances of the other species showed no systematic changes.

Our repeat surveys gave us an opportunity to do more than see what had, or had not, changed. Despite variation in bird abundances among years and sites, the statistical models we developed using our earlier data from 1977–1982 had done a good job of predicting bird abundances based on habitat features. Could we now use the new habitat measures from 1997 as inputs to those older models to predict bird abundances in 1997? In other words, did the statistical relationships between habitat variables and bird abundances that we documented from 1977–1982 still hold in 1997?

Generally speaking, the answer was "no." Of the fourteen bird species we encountered during our surveys in 1997, the models accurately predicted abundances of only Brewer's and black-throated sparrows. The problem, it seemed, was variation. In one-third of our surveys the abundance of a species in 1997 was outside the range of values we had recorded for a site in the earlier six-year period. Moreover, there was no clear pattern to which sites were lower or higher for which species. For example, abundances of sage and Brewer's sparrows in 1997 were both at the high end of the historical range for one site (Follyflat) but at another site (Catlow Valley) the abundance of Brewer's sparrows matched the historical high while sage sparrows were even less abundant than the lowest historical extreme.

The variation in abundances in 1997 eroded the predictive capacity of the models. The variability of the bird populations and the loose matching with habitat features were consistent with our overall conclusions that shrubsteppe bird communities were not in equilibrium and habitat was not fully occupied. There was also a broader lesson. The failure of statistically "good" models to predict future bird-habitat relationships suggested to us that ecologists or resource managers might want to be cautious in using models based on past conditions to extrapolate their expectations or management decisions into the future in a variable environment, especially at a local scale of resolution.[10] This suggestion was not greeted enthusiastically by our colleagues, although now, with the effects of climate change accelerating, the message is even more relevant.

〜〜

Let's return to the question that originally prompted our studies of shrubsteppe bird communities: could we detect evidence of competition among species, despite all this variation? Two of the shrubsteppe bird species, sage sparrows and black-throated sparrows, are closely related congeners that might be especially likely to compete. Although black-throats are more common than sage sparrows in the hotter, more arid deserts of the Southwest, they co-occurred with sage sparrows at five of our study sites in the northern Great Basin.[11] Densities of the two species varied inversely among those sites, suggesting competition. At the one site where both species were abundant, however, they occupied overlapping territories and seemed to ignore each other.[12] Although the differences in habitats and abundances of the two species among the study sites might reflect competitive interactions, it seemed to us more likely that the species simply exhibited different habitat preferences. They co-occurred where elements of habitats favored by both species were present. The different habitat preferences, of course, could be a product of competition sometime in the past that led the species to diverge in their habitat niches— a "ghost of competition past."[13] But we can only speculate on the conditions that might have prompted competition in the past, so this hypothesis is inherently untestable.

What about morphology and diets? As at our grassland sites, each shrubsteppe site contained several small species with low bill-size ratios, separated from a larger species by a considerable gap.[14] Bill sizes and body masses were unrelated to prey sizes. Overlap in prey sizes and diet composition among the species at a site was generally high. John Rotenberry found much the same results in his more intensive studies at ALE as part of his doctoral dissertation research. Although diets changed seasonally and between years, different species sampled at the same time were eating much the same prey, while birds of the same species sampled at different times were eating different things.[15] Overall, the birds were apparently foraging opportunistically on what was available at a particular time and responding to variations in prey availability in similar ways.

⁀⌒)

Collectively, the results of our studies of shrubsteppe bird communities paralleled those from the grasslands. The bird species varied in abundance independently of each other and showed little evidence of closely following yearly variations in vegetation structure or environmental conditions. Diets reflected opportunistic foraging and were unrelated to morphological differences among the species. Resources did not seem to be limiting (at least at the time of our studies) and available habitat was not fully occupied, leading to local reshufflings of territories and population density variations. In short, the patterns of niche differentiation expected from the competition paradigm failed to appear.

I had continued to cast my studies of bird communities in grasslands and shrubsteppe in the context of the competition paradigm, probably longer than I should have. This is what happens when a paradigm comes to dominate thinking in a science. You stick with the expectations of the paradigm longer than may be warranted, until the weight of the accumulated anomalies becomes too much to bear. It's partly human nature to want to stick with an appealing theory, particularly one that has wide acceptance in your field.

But at some point, you must open your mind to other possibilities. It took a while, but eventually I began to pursue an alternative, nonequilibrium view, one that held that ecological communities and systems are perhaps infrequently in the equilibrium state predicted by theory, especially in variable environments.[16]

11

CHALLENGING THE PARADIGM

Our findings from the grasslands and shrubsteppe were anomalies in the context of the prevailing competition paradigm. The ecological-crunch hypothesis challenged the underlying assumptions of the paradigm. This did not sit well with many of our ecological peers. We were called "exception ferreters" who sought to be contrarian rather than contribute to the advancement and strengthening of the paradigm.

Other ecologists were quick to offer explanations for why our results should be disregarded. Even though we had measured many habitat variables, perhaps we had measured the wrong things. They claimed we had not considered the critical niche dimensions, those whose variations the species tracked and over which they competed. Because the variety of habitat features that could be considered is virtually endless, however, this argument could be used to preserve the paradigm against almost any anomalous observations.

Or perhaps we looked at the wrong time. If the species we studied were limited by competition on their wintering grounds, for example, they might find resources to be superabundant on the breeding grounds. This would erode the expected patterns of niche displacement and preclude a close fit between morphology and diet. If the ecological crunch occurred in winter, our summertime results and interpretations would not be good tests of competition theory. But many of the studies that seemed to support the paradigm considered the niche relationships of co-occurring species on the breeding grounds. Winter season limitation was called into play only when anomalous results required an explanation.

Or perhaps we were focusing our efforts at the wrong scale. We did find consistent ecological differences among species at broad, regional scales, but they were overwhelmed by variation at the local scale of our study plots. But this is the scale where we should expect competitive interactions among individuals to be most intense.

Finally, some ecologists accepted our observations and interpretations but argued that the grassland and shrubsteppe bird communities, which contained only a few species, were "not typical" of bird communities in general and therefore were not good models that could be widely applied. Grasslands and shrubsteppe were "weird" and our anomalous observations could safely be ignored.[1]

I was not swayed by any of these arguments. But I eventually decided to shift my attention from digging deeper into bird-community organization to follow other pathways and address other questions. To understand why, it will help to delve a bit deeper into the debates that emerged during the 1970s and 1980s about the role of competition in structuring ecological communities.

The thesis that ecological communities were organized by competitive interactions that led to equilibrium conditions became widely accepted among ecologists in the 1960s and early 1970s. Anomalous observations such as ours were often discounted, sometimes quite vigorously. But ours were not the only studies that failed to confirm the predictions of the competition paradigm. Anomalies continued to accumulate, and the logical and methodological structure of competition studies was increasingly questioned. However, this did not lead to crisis in the Kuhnian sense, much less to a "revolution" and replacement by a new and better paradigm. Rather, as challenges mounted, there emerged three views of community organization that differed in the importance accorded species interactions and the prevalence of equilibrium conditions.

First, the prevailing competition paradigm held that competitive interactions among species led to assemblages that were in equilibrium at most times, in most places. The equilibrium argument was based to some extent on the foundations that Robert MacArthur and others had developed, which sought equilibrium solutions to mathematical models. It also reflected a broader belief in the balance of nature that had characterized Western thought for millennia. This approach influenced how ecological studies would be designed. If ecological systems were in a stable equilibrium, then conditions over several years should be similar and short-term, "snapshot" studies should suffice to reveal the patterns produced by competitive interactions.

It was obvious, however, that environmental conditions and interactions among species often varied over time. This prompted a second view, one that proposed that ecological communities were in a dynamic rather than a static equilibrium.[2] Resource availability changed as environments varied but species responded closely to these variations by continuously adjusting their competitive interactions. Community patterns of niche displacement and overlap might vary in time, but the changes reflected a changing competitive environment. Competition was always tuning species' responses, but it was expressed in different ways at different times.

A third view suggested instead that ecological systems in a variable environment would often be out of tune with resources and competitive interactions would occur only occasionally. Time lags in the responses of populations to resource conditions would create a "tracking inertia" that would complicate efforts to document competitive interactions. Long-term studies would be needed to reveal competition and niche displacement when they did occur. This "nonequilibrium" view became the dominant challenge to the competition paradigm during the late 1970s and 1980s. It accorded with the ecological-crunch hypothesis. It was what I felt was going on in the grasslands and shrubsteppe bird communities that I studied.

Competition theory suggested that species interactions would produce assemblages of species showing clear differences in such things as habitat occupancy, diet, or morphology—the patterns I had been looking for in my grassland and shrubsteppe studies. As challenges to the competition paradigm grew, some ecologists began to ask whether such patterns might also emerge if communities were assembled by chance alone, with no interactions among species. In other words, could the patterns used to support the competition paradigm arise as well if a local community was assembled by drawing species randomly from a regional species pool? This "null model" approach generated considerable controversy, in part because it was linked to the falsificationist philosophy espoused by Sir Karl Popper.[3] According to this philosophy, one should first falsify a null hypothesis of no interactions (i.e., random assembly) before moving on to test hypotheses based on processes such as species interactions. According to Popper, science advanced by falsifying one hypothesis after another, leaving standing only those hypotheses that could withstand multiple attempts at falsification.

Adherents to the competition paradigm were incensed by such suggestions. They argued, for example, that assembling a "community" by randomly picking species from a larger species pool would submerge the true relationships among a few pairs of species that did in fact compete in a mass of irrelevant detail from other species that one would never expect to compete. Philosophically, they took issue with the basic Popperian approach. Why would you want to go to considerable effort to determine that a proposition was false? This would only show what wasn't, not what was. Where would that leave you?

These arguments, like many in ecology, sometimes degenerated into polarized "either this or that" debates. At the extremes, the "equilibrium" and "nonequilibrium" views of communities posited quite different things. Equilibrium communities were tightly coupled by species interactions, particularly competition; resources were limiting and fully used; the species present made optimal use of the resources in order to coexist; and patterns of community organization were tight and predictable. Nonequilibrium communities had the opposite characteristics.

Because the variable environment/ecological-crunch ideas I had proposed most closely aligned with the nonequilibrium view, I was called upon to promote and defend that view. And I did so (while avoiding being drawn into the null model debates).[4] But I felt the reality lay somewhere between the two extremes. Some communities might be resource limited, competitive, and in equilibrium much of the time, others only rarely so. Some patterns might be evident at some scales or in some places but not others. Some members of a community might be competitors and others not, and those which were or were not might change from place to place or time to time. Thinking of ecological communities as either equilibrial or nonequilibrial created a false dichotomy. Natural communities might best be viewed as arrayed along a spectrum between equilibrium and nonequilibrium, their positions varying in time and space as environmental conditions varied.

This view of reality, of course, meant that it would be much more difficult than I first thought to understand how (or even whether) bird communities were organized. It also meant that the broader debate about the role of competition in structuring communities was not likely to be easily resolved. More detailed observations and field experiments would be required.

It turned out that a Belgian colleague, André Dhondt, had been conducting long-term studies of competition between great and blue tits in Belgium. André, who had now moved to Cornell University, spent a sabbatical semester with me in 2002. Among many things, we discussed my challenges to the competition paradigm. André's experiments had provided clear evidence of competition. We agreed that competition could be a major force structuring bird communities in some situations but not others. Documenting its

importance required well-designed experiments or detailed observations, not just assumptions. A decade later, André published a
definitive treatment of the evidence about interspecific competition
in birds.[5] Then, to bring things full circle, my son David conducted
his PhD studies on interactions between spotted and barred owls,
providing clear evidence of how the impacts of competition contributed to the deteriorating conservation status of the spotted owl in
northwestern forests.[6]

Evaluating the role of interspecific competition and unraveling the
forces that structure bird communities had dominated my research
for two decades. It was unlikely that the debates over the role of
competition in structuring ecological communities would be
resolved to everyone's satisfaction. I decided it was time to move on
to other questions and explore other pathways. But I was not ready
to leave these issues entirely just yet. I decided to distill my thinking into a book, *The Ecology of Bird Communities*, which was published
in 1989.[7] Actually, "distill" is not accurate; the book turned out to
be two volumes with a total of 855 pages. Apparently, I had a lot
to say, much of it casting doubt on the ubiquity of competition in
bird communities.

Reflecting now on my role in the "competition debates," I can see
an important lesson that was not apparent to me at the time. If
you're going to challenge an established paradigm (or even an established body of theory), be sure that your studies can stand up to
intense scrutiny. Scientists should do this anyway, of course, but
when a paradigm and normal science hold sway, anomalous findings
that challenge the very foundation of the paradigm may be more
critically evaluated than those that don't. And it's probably best not
to challenge an established paradigm when you're just starting your
career. To be successful in ecology (as in any science) you'll need to
join the scientific culture, and you don't want to begin by alienat-
ing your peers. But you must be true to your data. If your findings

really are anomalies that threaten the veracity of the paradigm, you should stick by them. Just be diplomatic and don't let your challenges become personal. Unfortunately, attacks on the professional integrity of some participants in the null model debates confounded efforts to gauge the value of this approach and prolonged the debates.

12

CHANGING PLACES
Pathways and Practicalities

I suggested in the Introduction that several forces act to direct a scientist onto pathways as a career develops. One of these forces includes *life's practicalities*—the personal events and "other things" that make up a life. Changes in jobs and locations are part of this, of course, but other personal changes can also have an overpowering effect on how pathways develop. Often these changes are not anticipated so you can't really plan how they will affect your pathways. Instead, you must decide which pathways to follow in the context of life's changes.

Time, then, to recount two life changes that had profound effects on the pathways I would follow.

～～

Like many scientists, I was an academic—a university professor. By 1977 the shrubsteppe research was well underway, yielding insights to challenge the ubiquity of the competition/equilibrium paradigm. I was comfortable in my faculty position at Oregon State. I had just finished a stint as Acting Chairman of the Zoology Department and was looking forward to resuming a normal life. Our family had settled into living in Corvallis. I had not given any thought to seeking a position elsewhere.

But that fall I got a call from my old collecting buddy David Ligon, who was now on the faculty at the University of New Mexico (UNM). The UNM Biology Department was pushing for approval to recruit a senior ecologist. David wondered if I could send him a copy of my curriculum vitae to show their dean the kind of person they had in mind. I was happy to help out, so I sent David my CV and promptly forgot about it.

Several months went by. Then one day in early spring, I got a call from the UNM Biology Department chair, asking when I could come for an interview. "For what?" I asked, having forgotten all about this. It turned out that David had included my CV in the pool of applicants for the senior ecologist position. I was one of three finalists for a position that I didn't know I had applied for.

At this time, I wasn't particularly interested in leaving Oregon State. NSF had agreed to continue funding my shrubsteppe work (perhaps in the hope that with more data my results would line up with the paradigm). But I thought it would be fun to visit David and see what New Mexico had to offer. So I went for the interview. The discussions following my talk about the mismatch between our findings and the competition paradigm were wide-ranging and stimulating. I was impressed by the diversity of interests in the department. And New Mexico, with its mixing of many cultures, was quite different from Oregon. At the end of my visit, I spent some time with David in the Chihuahuan Desert south of Albuquerque. The openness and dryness of the New Mexico landscapes appealed to me. "We could live here," I thought.

But then I returned to Oregon and didn't think much more about it. A week later, however, I got a call offering me the position—at a substantially greater salary. Even then I was reluctant to leave Oregon, but I dutifully played the academic game of dangling the offer in front of my dean to see what Oregon State might do as a counteroffer. Not much, it turned out. Oregon State had long banked on the allure of Oregon, the livability of Corvallis, and individual inertia, rather than financial rewards to retain faculty. But I had seen enough of New Mexico and the university to find the prospects exciting, and the thought of again being colleagues with David was compelling.

I accepted their offer. We sold our house in Corvallis and made a quick trip to Albuquerque to find a new place to live and make sure our kids were registered for school. I broke the news to my graduate students, offering them the option of coming along or finishing their degrees at Oregon State with another professor. Most followed—more, I think, because of the excitement of living in New Mexico than of any sense of allegiance to me. But I was delighted to have them come along.

Several of the students took advantage of the opportunities for field research in the deserts of New Mexico. However, John Rotenberry and I were committed to continuing the shrubsteppe work begun at Cabin Lake for another year or two. It turned out to be another five years. My pathway continued to run through the shrubsteppe. The change in location didn't immediately affect my research pathway. But the new colleagues at UNM were to redirect my interests later on, as we shall see.

⁓

The pathways you follow in a career as a scientist aren't all determined by science. Sure, interests, training, opportunities, colleagues, success in publishing papers, and getting research grants all influence which pathways are followed, and which aren't. Obviously, location also plays a big role. But scientists are people, and events in their personal life have a huge influence on what they do and where they go.

So it was with me.

When I left Oklahoma for Wisconsin, Martha and I had decided it would be a good time to get married. So we did. We had two children, Ann and David. During our time in Oregon, however, the marriage began to show strains, and shortly after the move to New Mexico it fell apart. We agreed to a divorce. Martha went her way, while Ann and David stayed with me in Albuquerque. I sold the house, moved to a smaller place, took care of the kids, and spent the evenings reading philosophy. Then things changed.

Several years before, back in Oregon, Bea Van Horne had joined my group of graduate students to pursue a master's degree. She completed a project on small mammals in Alaska and decided to stay on for a PhD. She obtained support from the U.S. Forest Service for a larger project on the responses of small mammals to forest succession following clearcut logging in southeastern Alaska rainforests. When I made my move to New Mexico, Bea decided to move as well because that seemed the best route to complete her degree and progress professionally.

During her last PhD field season in 1980, I had an opportunity to visit Bea at her study site in Alaska to check up on how she was doing. At that time, I was involved in a project to assess the potential effects on seabirds of oil development on the Alaskan continental shelf (more on that in chapter 20). The postdoctoral associate on the project, Glenn Ford, and I were asked to come to Alaska to report on our progress. Glenn had studied small mammal ecology for his PhD, so the two of us added a side trip to Bea's field site. We few in by floatplane to her base at a U.S. Forest Service camp on Prince of Wales Island and spent two days touring her study areas, checking her small-mammal traps, and talking ecology.

Somehow, sometime during that visit, Bea and I fell in love. There was no warning. It just happened. Our previous relationship had been entirely academic. I was Bea's professor and mentor; she was one of my students. But now, suddenly, everything had changed. I might have been content to continue reading philosophy in the evenings, but no longer. Instead, I wrote bad poetry to Bea. Neither of us could go back to the way things were before. Talk about tipping points—our lives were changed forever. Our pathways were now intertwined, and they have remained so for over four decades. Bea has been a collaborator and critic, someone with whom I can share my wildest ideas, whose understanding of ecology has been honed by field experience. She's been a steady companion and a mother to our children, a fellow traveler on the pathways of ecology and conservation. My science, and my life, has been richer for it.

Bea and I continued to see each other after she returned to Albuquerque in the fall. Bea completed her PhD at the end of the fall term in 1981 and stayed on at New Mexico conducting her own research. After my divorce was finalized, she joined Ann, David, and me to form a household—a family. We were married in the spring of 1983. David Ligon was my best man.

We settled into our shared life in Albuquerque. Bea continued her research, conducting a large study of the impacts of military training on the population dynamics of ground squirrels that were the primary prey of several raptor species in the Snake River Birds of Prey National Conservation Area in Idaho. I continued working in the Oregon shrubsteppe and began planning for a sabbatical leave in 1984–1985. But we both realized that this arrangement was untenable. Bea had developed her own professional identity and reputation as an ecologist. To continue as a research associate on grants that she generated would not be professionally, intellectually, or personally satisfying and would waste what she could contribute as a scientist and teacher. She deserved her own faculty position and the opportunity and resources to forge her own programs.

After we returned from the sabbatical in Australia (more on that in chapter 13), we realized that prospects at UNM were probably slim. We liked living in New Mexico, I was developing new collaborations with other faculty in the UNM Biology Department, and Bea could probably keep generating research funding. But having gotten her PhD there, she might forever be regarded as a graduate student. Although staying in Albuquerque had a lot to offer, apparently that did not include a faculty position for Bea. We began looking for a place where both of us could have full-time faculty positions—ideally, someplace in the West. We each started exploring possibilities, thinking that if one of us got an offer we might be able to negotiate a position for the other.

Then, quite unexpectedly, I got a call from a colleague at Colorado State University (CSU). The university wanted to enhance their ecology programs and the CSU Biology Department was recruiting a senior ecologist. Would I be interested in applying? We thought it over for perhaps five minutes. It seemed to be exactly what we were looking for, *if* another faculty position could be arranged. We'd cross that bridge if we came to it.

I interviewed for the position and explained our situation. Not only did they offer me the job, but they also wanted to interview Bea. She interviewed, and they offered her a faculty position in the CSU Biology Department as well. Offers in hand, we wanted to see what New Mexico could do as a counteroffer. Faced with the prospect of losing both of us, it took only a few minutes for the biology chairman to convince his dean to come up with a full-time position for Bea. But it was too late; we'd already made the decision. Our future intertwined pathways ran through Colorado.

And so, in 1986, I—we—moved again. It was all very serendipitous. I didn't plan on falling in love, getting together with Bea, or moving to Colorado. But life happens. As I said, career pathways involve more than science.

13

TESTING THE PARADIGM

Are Australian Bird Communities Different?

Paradigms, once established, can have a tenacious grip on a science. They focus the questions asked but also constrain how the questions are answered. Thus, even though I had used our findings in grasslands and shrubsteppe to challenge the competition paradigm, I couldn't entirely shake free. I had yet one more opportunity to delve into the patterns of bird community organization.

One of the greatest benefits of being a university professor is the sabbatical leave. Sabbaticals allow you to experience a new setting, interact with new colleagues, explore new research possibilities, develop new skills, and just spend time thinking and writing. No matter how much you enjoy university life, over the years it can wear you down. I was due for a sabbatical from New Mexico in 1984–1985, so I started planning well before that. I needed to decide what to do and where to do it.

The "where" part was easy. I suspect that most North American ecologists at some point yearn to spend time in Australia, simply because its flora and fauna are so different from what we're used to. We can imagine that ecology in Australia plays by the same rules that ecologists have derived for the Northern Hemisphere, but that assumption should be put to a test. I wanted to spend my sabbatical in Australia.

Doing so would require funding. The university would cover half my salary while I was away, but I needed to generate the other half and what I would need for travel and research expenses. I decided to apply for a Fulbright Senior Scholar Award. The Australian-American Fulbright Commission offered awards to foster bilateral academic and cultural exchanges for "research or study at an Australian institution, experiencing life in Australia, and bringing back their knowledge and experience to share with their communities in the U.S." It seemed perfect. However, the application required that I specify the "what to do" part of my plan, which led me back into the grip of the competition paradigm.

The notion of ecological convergence had been part of evolutionary thinking since at least the time of Darwin. Even though they differ in their evolutionary history, species living on different continents might be expected to adapt to similar environments in similar ways. They would converge morphologically and ecologically. Competition might lead to convergence in habitat use, behavior, or life history of species and similarities in community organization.

Shrub deserts in Australia would seem to be ideal places to test such ideas. Despite the differences in seasonality, geology, evolutionary history, and climate, Australian shrub deserts are strikingly similar to those in North America.[1] An obvious question, then, is whether the birds and bird communities in North American and Australian shrub deserts are ecologically or morphologically convergent. At least that was the question I posed in my Fulbright application. It was firmly anchored in the competition paradigm.

The Fulbright process required that I be sponsored by an Australian scientist. Harry Recher, an ornithologist at the Australian Museum who had worked with Robert MacArthur, seemed an ideal choice. Harry agreed, although he pointed out that we both knew I no longer accepted the competition-based explanation of convergence. No matter. I was awarded the Fulbright. David and I were off to Australia for a year (Bea, who had her own research projects going, would join us later and Ann was now in college studying to become an artist).

Proposing a research project was the easy part. I now needed to deal with the realities. To determine whether and how birds in Australian shrub deserts converged ecologically with their counterparts in the sagebrush shrubsteppe would require fieldwork. Indeed, that's why I had proposed such a study—to experience the wide-open spaces of the Australian outback. Where could I find a match for Cabin Lake and our other Great Basin sites? I talked with Harry and colleagues at the University of Sydney, where I was based. Someone suggested that the University of New South Wales had a field station in the outback that might be suitable. We decided to head there to check it out.

Distances in the Australian outback are even greater than those in the American West. Places are far apart, and it takes a long time to get anywhere. The Fowlers Gap Arid Lands Research Station was some 100 km north of Broken Hill, which was more than 1,100 km from Sydney. It took nearly two days to get there. The research station had been a sheep station that went bust during one of the interminable Australian droughts and had been acquired by the university. It was now largely used to study sheep production under the trying conditions of the outback. There was a cottage where we could stay during the fieldwork—pure luxury compared to our camp at Cabin Lake (or our middle-of-the-road campsites during the rapid surveys). The station even provided all the mutton we could eat while we were there. We discovered that there are many ways to make interesting meals with mutton.

The station lands around Fowlers Gap included several very large paddocks, none of which was currently being grazed (sheep flocks had been culled during a lengthy drought). The paddocks were dominated by shrubsteppe that looked remarkably similar to our Great Basin sites. It had the same feel. If you closed your eyes, you could imagine that you were in the Oregon shrubsteppe—until you opened them to see a big red kangaroo staring you down. The lands at

Fowlers Gap provided plenty of room for me to set up study plots wherever I liked. And it was still early spring in Australia, so the birds were just beginning their breeding season. Fowlers Gap would be an ideal place to look for evidence of ecological convergence. We ended up spending several months living there.

EPISODE: LOOKING FOR CONVERGENCE IN THE OUTBACK

To allow similarities and differences in the bird communities of the Great Basin and Fowlers Gap to show through I needed to minimize other confounding differences, so I used the same study design and observational methods I'd been using in the Oregon shrubsteppe. I set up four study plots in different paddocks; measured habitat features, mapped territories, and recorded habitat use as I had been doing for years; and analyzed the data in the same way.[2]

The habitat measures confirmed my initial impressions. Although the shrub species in Australia were different from those in North America, the habitats were structurally indistinguishable. Thus, the habitat setting for ecological convergence among the birds was clearly there.

The birds, however, didn't play along. The number of bird species on the plots was similar (averaging 6.3 in North America versus 5.5 in Australia). Densities of breeding birds in Australia, however, were only half those in North America (158 vs. 301 individuals km^{-2}). Other investigators had found much the same pattern of differences in densities in comparisons of desert mammal and lizard faunas between the continents.[3] I'll offer a possible explanation in a few paragraphs.

To assess whether species in the North American and Australian shrubsteppe converged ecologically, I assembled information on multiple ecological and life-history attributes of species from conversations with Australian ornithologists and my own observations. You wouldn't expect every bird species on a plot in one continent to have a close counterpart in a plot in the other, but if convergence is common there should be several pairs of species that share attributes. Only one close pairing emerged out of seventy-two possible pairings of a North American with an Australian species: western meadowlarks shared many attributes with brown songlarks.

So there was little close matching between individual pairs of species on the two continents. Perhaps the groups of Australian and North American bird species that occurred on my plots collectively shared similar life-history adaptations to the similar shrub-desert habitats. Here also, there was little evidence of convergence. In comparison to their North American counterparts, the Australian species differed in several basic life-history features. As a group they had longer breeding seasons, more broods per year, were more likely to form social groups, and were sedentary or nomadic rather than seasonal migrants.

What about morphology? If morphology reflects ecology, then one might expect species in similar ecological settings to be similar morphologically. If species in a plot are spaced out morphologically (perhaps by competition), then one Australian species might be more like a North American counterpart than another Australian species on the plot. To test this idea, I used several morphological measures—things like body size, bill length, and wing length. For the North American species, I could use measurements of birds on our study plots. For the Australian species I spent a week measuring specimens in the collections of the Australian Museum in Sydney.

There were more morphological pairings of species at Fowlers Gap with those in the Oregon shrubsteppe than one would expect by chance alone.[4] Richard's pipit paired with horned lark, southern whiteface with Brewer's sparrow, and (again) brown songlark with western meadowlark. Although these pairings might in fact reflect similar adaptive responses to structurally similar habitats, the assumption that they were consequences of competition between species seemed much more tenuous.

One reason is that the structural similarities of the vegetation at Fowlers Gap and the Great Basin were only part of the picture. Soils in the Australian interior are old, weathered, and nutrient-poor in comparison to those in the sagebrush shrubsteppe. Productivity is lower, which could account for both the wider ecomorphological spacing of the Australian species and the substantially lower breeding densities I recorded at Fowlers Gap. And the climates are quite different. Although annual precipitation averaged over many years was similar in the two areas, it was much more variable, and droughts lasted much longer and were much more widespread in Australia.

There's a saying in the outback that "rivers run dry or 10 feet high." When I was working at Fowlers Gap it hadn't rained for eighteen months. The pastoralist down the road offered that it was "a bit of a dry" and he was thinking of cutting back his sheep herd. A week after our last trip to Fowlers Gap it rained over 20 cm in two days. Australia is flat, so climate extremes are likely to be widespread. Such conditions favor sedentariness or nomadism in the breeding birds. Where we worked in the northern Great Basin, winters are cold and the basin-range topography limits the spatial extent of climatic extremes; conditions are likely to be different a short distance away. All the birds that bred on our North American plots were seasonal migrants.

Overall, the differences in the bird species and communities in my Australian and North American study areas were far greater than the similarities. Although the vegetation structure was remarkably similar, everything else was different. My expectation that it might be a good setting to look for ecological convergence was probably naïve. But I certainly learned a lot.

~⁀◞

The year in Australia, of course, wasn't all science and research. That got us to Australia, but then we wanted to experience the different place, different environment, and different people. That's a large part of what the Fulbright program is all about.

David and I arrived in Sydney in late August 1984—early spring in Australia. I rented a house north of the city in St. Ives, bought a car that I thought could withstand the outback roads, and registered David for school. His reaction to the idea of wearing a school uniform every day was somewhat mollified by the realization that he'd be getting two summer breaks in a single year.

We began to enjoy the urban birdlife. We were awakened in the morning by a group of kookaburras chorusing in our backyard. Rainbow lorikeets fed on nectar from flowering shrubs in our front yard, and raucous flocks of galahs paraded on our lawn and tried to perch on the utility wires, struggling to remain upright but usually ending up hanging upside down. An Australian magpie frequently attacked us as David and I walked to his school. There was a large area of bush a block away that provided habitat for a diversity of birds—cockatoos and corellas, whistlers, fantails, fairy wrens, scrub wrens, thornbills, honeyeaters; even a superb lyrebird if you knew where to look. It was all very different from our place in Oregon. I hadn't kept a bird list since I was a boy birdwatcher, but I started one now.

While David spent his days in school, learning (among other things) the intricacies of the strange game called cricket, I began writing the book that I mentioned earlier, summarizing my thoughts on bird communities. The project ended up taking much of the time we were in St. Ives and was much more than I had envisioned. I felt compelled to review and comment on many studies and many topics. By the time we returned to Oregon I had completed a draft of *The Ecology of Bird Communities*.

Bea joined us in December. David was on summer break, and I was more than ready for a break from writing. We spent most of December and January at Fowlers Gap. We were there for Christmas. We were used to Christmas being in winter, often with snow (the "White Christmas" of song). At Fowlers Gap it was summer and hot. We fashioned a Christmas tree of sorts from a branch of *Casuarina* and decorated it with a star made of aluminum foil. We had a fine Christmas dinner of barbequed mutton. Other researchers had left the station for the holiday, but the staff remained. There was even a Santa for the kids—one of the mechanics outfitted with a red shirt, shorts, and a sheepswool beard—who rode in on a motorcycle, bellowing "Ho, ho, ho" with a heavy Aussie accent. We felt right at home.

We also explored some other places, of course. Not as many as we had planned—Australia is a *big* place. Two excursions stand out.

One was to O'Reilly's Rainforest Retreat in southern Queensland. O'Reilly's is a hot spot for birding. We took advantage of the trails leading off into the bush and rainforest of Lamington National Park, adding to our bird list and experiencing habitats we'd never seen before. On one of our hikes through the forest we even witnessed an unusual example of avian tool use. Bea and I heard a loud, melodious song punctuated by banging sounds. We crept through the brush to find an Albert's lyrebird on its display arena. The bird had created a musical instrument to supplement the attractiveness of its song: it repeatedly pressed its foot down on a branch connected to a hollow log, which produced a resonate, percussion-like bang. Although we watched for nearly an hour, no female showed up. Perhaps the song was too strange, but the male kept at it. He was still orchestrating his song the next day.

One shouldn't spend time in Australia without visiting the Great Barrier Reef. To fully appreciate the reef and its glorious biodiversity, however, you really need to scuba dive. None of us had dived before, so we dutifully signed up for scuba lessons at a dive shop in St. Ives. Bea and David mastered it easily. I found it more challenging. I really don't like water—it's one of the reasons I like to work in deserts. The thought of being *underwater* for any length of time didn't appeal to me. But there was no other way to see the reef, so I stuck with it and eventually passed certification. We headed out to Heron Island.

And it was well worth it. The splendor of the reef was something to behold. I was so excited on our first dive that I used my forty-five-minute air supply in fifteen minutes. After I calmed down, I learned to float in one place just observing all that was going on—damselfish defending their tiny territories; parrotfish grazing on the coral; schools of needlefish; morays lurking in cavities in the coral; multiple species of butterfly fish milling about; cleaner wrasses tending to their clients; and the occasional reef shark cruising by. We did two dives a day for the week we were on Heron Island. Of course, there were other attractions on the island—rails, white-eyes, noddies, hatchling sea turtles—but nothing could compare with the world underwater.

After Bea returned to the United Staes to continue her research in March, David and I settled back into our routines, he in school and hanging out with friends playing at cricket or going to the beach to surf, me writing and assembling our observations from Fowlers Gap. Then, before we were ready, it was time to head home. Such is the nature of sabbaticals: a lot of effort to put one together, a hugely rewarding and productive experience, and over all too soon.

The time in Australia was a change in location, albeit a temporary one. How did it affect my pathways? It resurrected my interest in the organization of bird communities, but with the completion of my book that pathway finally came to an end. Rather than establish a new Aussie-centric pathway, however, I came away with perspectives that I would carry with me whatever pathways I would follow. The Australian shrub deserts reinforced my belief that ecological understanding demands a consideration of the effects of environmental variation in time and space. Australian species, communities, and ecosystems are different because the environmental setting and its history are different. I realized that the concepts and theories developed by ecologists working in the more benign and (relatively) stable environments of the Northern Hemisphere need to be closely scrutinized before they are applied uncritically to ecosystems elsewhere on Earth. The Australian ecologists that I met who had grown up studying their environments came to them with different expectations and asked different questions. Their perspectives resonated with me.

Chief among these was one of my ecological heroes, Charles Birch. In 1954, he and H. G. Andrewartha had published a book, *The Distribution and Abundance of Animals*.[5] They set forth in great detail their argument that the environment determined the size of populations and their distributions. This "density-independent" view contrasted sharply with the prevailing position of American and European ecologists, that populations were self-regulated by density-dependent competition for limited resources—the competition paradigm. These differing perspectives were being vigorously debated when I was in graduate school. Although I had initially embraced the competition paradigm, the results of my studies consistently failed to match its predictions. By the time I was in Australia I had come to advocate the importance of the environmental factors (especially weather) promoted by Andrewartha and Birch. Charles and I found much to talk about. He also thought deeply and wrote much about the intersection of science with religion, so our discussions often wandered into philosophy.[6]

~

For an ecologist, there is no substitute for spending time in another part of the world. You have an opportunity to consider everything you've learned about nature in a new light. The species, communities, and habitats are different, challenging the beliefs, assumptions, and paradigms you've come to regard as essential truths about how nature works. And the ecologists you encounter often have a different, fresh perspective on what you see and study. Ecologists who don't travel widely are missing something vital.

14

SHIFTING DIRECTIONS IN THE SHRUBSTEPPE

My pathway in North American shrubsteppe studies had begun to shift in different directions even before I spent a sabbatical "down under." Let me summarize what brought me to that point.

After I completed my PhD in 1966, I continued down the pathway investigating the ecology of bird communities. I sought to show how bird communities in simple settings could advance our understanding of the role of competition in structuring communities. My work was couched in the currency of the competition paradigm. Although it soon became apparent that our findings didn't fit—bird communities in grasslands and shrubsteppe seemed to be anomalies. Such is the power of a paradigm, however, that I persisted in my attempts to reconcile our work with the paradigm through moves, changes in jobs, and changes in the study ecosystem. My sabbatical in Australia was one more attempt to validate the paradigm. The paradigm endured because it was not obviously wrong, just not as broadly applicable as was once thought. It was still attractive to many ecologists and remained part of the scientific culture of ecology. This culture—peer pressure—was a powerful force acting to keep me going on this pathway.

But no one likes to have their work dismissed as irrelevant anomalies. I began to scramble for alternative explanations. Over time, I came to realize that climatic variation might lead to conditions in which the core assumptions of the paradigm—resource limitation and equilibrium—might not always hold. My pathway shifted. I was no longer searching for patterns in the organization of bird communities that might align with the competition paradigm. I asked different questions, ones more directly related to environmental variation and nonequilibrium and, in the shrubsteppe, focused on behavior. How did the activities of different species relate to habitat structure? How did individuals partition their time among various activities (their activity budgets)? How did activity budgets differ among coexisting species? Did they vary between years or locations? Why? These were some of the same questions I had asked in my grassland study in Wisconsin.

John Rotenberry and I were now full-fledged collaborators on the research. We emphasized sage and Brewer's sparrows, the dominant species in sagebrush habitats. Our methods were simple but sometimes tedious. One of our field crew would select a bird for observation and follow it for up to half an hour, recording what it was doing and where it was at twenty-second intervals. We did this, off and on, for three breeding seasons, producing mountains of data.

What we mostly saw was more variability. Activity budgeting and habitat use by both species varied between study plots at a site, between sites, and between years, and the variations were unrelated to differences in vegetation coverage or other environmental features at the sites.[1] At Cabin Lake, where our observations were most intensive, the two species occupied most of the study plots, so territories of both species overlapped and contained similar coverage of sagebrush, bare ground, and grasses. Overall, the similarities between the two species were much more obvious than any differences.

We were particularly interested in foraging behavior because of its close linkage to potential food limitation. The distribution of sagebrush as discrete patches (individual shrubs) also provided an opportunity to relate foraging behavior more specifically to patch use.[2] Both species spent most of their foraging activity in sagebrush, while seemingly avoiding patches dominated by rabbitbrush (*Chrysothamnus* species). Patches of sagebrush at Cabin Lake differed slightly (but significantly) in size and foliage growth among years, likely in response to two abnormally wet springs. Both sparrow species responded to the yearly changes in the same way, spending more time foraging in larger, more vigorous shrubs and more time in a shrub if they had to travel farther from another shrub to get there. The obvious question was whether this nonrandom use of shrub patches was related to food availability.

It turned out that it was. We used a "D-Vac" (a modified vacuum cleaner) to collect samples of arthropods from a random set of shrub patches and used the information from our previous diet studies to categorize which insects were bird prey. Sure enough, the larger, more vigorous sagebrush patches where the birds foraged more often contained significantly more bird prey (both numbers and biomass). The birds seemed to respond opportunistically to variations in prey availability among shrubs and years. This observation, however, didn't tell us whether food was limiting to the birds, in which case we might expect yearly differences in patch use to be related to reproductive success.

Overall, variations in reproduction of the two species were unrelated to habitat use. We recorded the fates of over one hundred nests of sage and Brewer's sparrows over two to three years at the ALE, Cabin Lake, and Star Creek sites.[3] Clutch sizes did not differ among the sites or between the species (both laid on average three eggs per nesting attempt). Success in hatching eggs and fledging young, however, differed markedly among the sites: success was high at Cabin Lake, intermediate at ALE, and low at Star Creek. There was a simple explanation: snakes. As I took some delight in noting before, there were no snakes at Cabin Lake. But they were plentiful at Star Creek, so much so that in one year all of the Brewer's sparrow nests we found suffered snake predation before any eggs hatched.

Variation in precipitation also affected the birds, but in different ways for the two species.[4] Sage sparrow hatching success was greater in wetter years while clutch size remained unchanged; Brewer's sparrows responded to between-year variations by increasing clutch size rather than hatching success.

One of the most obvious behaviors of birds in the shrubsteppe (and birds in general) is singing. One of the joys of working in the shrubsteppe was hearing the hauntingly melodious songs of sage and Brewer's sparrows drifting across the sagebrush in the early morning light. When John and I were conducting our rapid surveys, these songs got us going in the mornings as we rolled out of our sleeping bags.

Brewer's sparrows sang longer songs than sage sparrows, often going on uninterrupted for a minute or more. Consequently, they spent more of their time singing than sage sparrows, and because they sang from the tops of sagebrush they centered more of their behavior on sagebrush.[5] Sage sparrows sang a relatively simple song of eight to ten notes strung together for a few seconds, followed by another short burst of song half a minute later.

As John and I visited our rapid survey sites I noticed that sage sparrow songs sounded different at different locations. At this time, in the mid-1970s, ornithologists were giving a lot of attention to patterns of geographic variation in birdsong. Several studies had described "song dialects," in which individuals in one location sang virtually identical songs that differed from the songs of another group of individuals some distance away, which were likewise similar to one another. Dialects were widely thought to develop when young birds learned the vocal idiosyncrasies of their fathers' or neighbors' songs, incorporated them into their own vocal repertoire, and subsequently remained in or returned to the same local area to establish breeding territories as adults. I began to wonder whether sage sparrows might also have vocal dialects.

EPISODE: THE SONG OF THE SAGE SPARROW

I decided to embark down a side pathway dealing with birdsong. Several of my students had studied bird songs, so the equipment necessary to record and analyze song patterns was available in my lab. It would be easy to record sparrow songs as John and I visited our network of rapid survey sites. It turned out, however, to be a longer diversion than I expected. I spent three years recording sage sparrow songs and another two years analyzing the structure of the songs.

So I added a birdsong component to our community surveys. I spent a morning at each of our sites walking our survey transect with a reel-to-reel tape recorder, a dynamic microphone, and a large and unwieldy parabolic reflector recording the songs of individuals and mapping their locations so I could determine song similarities among neighbors. Because we were working well off the beaten path, we never encountered anyone and I was spared having to explain why I was aiming a large parabolic reflector at something somewhere out in the sea of sagebrush.

Characterizing individual song patterns involved producing a sound spectrograph of a song, identifying individual note types within the song, assigning each note type a numeric code, and then converting the song into a code sequence. The code sequences could then be compared among individuals and analyzed statistically to quantify song similarities between individuals at a site and among sites. It was a laborious process; much of it can now be done digitally.

The patterns that emerged were inconsistent.[6] At some sites all the individuals I recorded sang essentially the same song—the notes, sequences, timing, and frequencies were nearly identical. At other sites, neighboring individuals had quite different song patterns. And at one site (Guano Valley), clusters of three to five neighbors sang similar songs, which differed markedly from the songs of three to five neighbors in an adjacent cluster. At a broader, between-site scale, songs of birds at some sites were quite similar to those at a nearby site, but in other cases songs of individuals at sites only 1–2 km apart were different from each other but were more similar to the songs of birds at different sites some distance away. Song similarity was generally unrelated to the distance between sites, habitat characteristics, or the local abundance of sage sparrows or other species.

Once again, the shrubsteppe had refused to give an unambiguous answer to what I thought was a simple question. Song variation in sage sparrows seemed to be all over the map (literally). Grasping for an explanation, the best I could muster was a suggestion that local populations of sage sparrows might differ in the return rate of adults to previous breeding territories. Where yearly turnover among the males holding territories was high, new males with different song patterns might fill the gaps, providing different models from which young birds could learn. The young birds, in turn, might disperse to fill gaps at other breeding locations. All of these factors could produce a haphazard mixing of song patterns within and among locations depending on the return rate of adults, the variety of songs to which young birds were exposed, and their subsequent dispersal. There was nothing resembling the dialects that had been reported for other species, reports that were often accompanied by speculations about the evolutionary or adaptive consequences of a dialect song structure in populations. I labeled the variations "epiphenomena" that were due largely to chance events and had little adaptive significance.

The idea of song dialects had become a mini-paradigm of sorts. My interpretation and the "epiphenomenon" label were rejected by many birdsong researchers. My observations were considered to be an anomaly, yet another manifestation of the weirdness of the shrubsteppe.

I had simply wanted to see if there was a pattern to the variations in sage sparrow songs that I heard. I got started on this because I was interested. But interest alone is not enough to change a career pathway. To pursue this pathway further I would need to get more deeply involved in exploring the details of birdsong ontogeny and learning. To do it right would entail aviary and laboratory studies. That was not a pathway I wanted to follow. There were other ways I might be able to wring some answers out of the shrubsteppe. Perhaps if I conducted experiments I could come up with more consistent and interpretable results. It seemed worth a try.

15

FINDING THINGS OUT

Field Experiments in the Shrubsteppe

Field experiments have long been an important part of an ecologist's toolbox. Of course, you can test a hypothesis (or story) that you've developed by comparing places with and without some important feature or by correlating natural environmental variations with the responses of individuals or communities. But this will take you only so far. At some point it may be more informative to conduct a field experiment, to manipulate some elements of a natural situation to see what happens. It's an important part of getting at underlying causes—finding things out.

John and I had spent considerable time in the shrubsteppe trying to figure out how birds responded to habitat structure. Our observations, it seemed, were plagued by multiple sources of variation—between plots, among sites, among years, and between species. Our behavioral studies had shown that sagebrush was heavily used by birds for foraging and singing. What would happen if we experimentally manipulated sagebrush availability?

An opportunity to address this question came along quite unexpectedly while John and I were conducting our rapid surveys to assess bird-community patterns. It actually wasn't a true experiment with controls and such. Rather, it was a habitat manipulation that we hadn't planned, that was conducted by someone else. The "someone else" was the U.S. Bureau of Land Management (BLM).

Guano Valley in southern Oregon was one of our favorite rapid-survey sites.[1] There was a vast expanse of sagebrush on the ancient lake bed of the valley floor, with the imposing scarp of a 250 m fault block rising in the background. We had already surveyed the birds there for three years when we drove down the dirt track to our study area and campsite in the summer of 1980. We looked about in disbelief. Sagebrush grows very slowly, but here the sagebrush had suddenly gotten bigger and greener.

It turned out that BLM had initiated "range improvement" over a large area that included our study plot. The sagebrush had been aerially sprayed with the herbicide 2,4-D that spring before the breeding birds arrived. The herbicide induced a spurt of plant growth, after which the shrubs would die. The dead shrubs would be removed the following fall and crested wheatgrass would be planted to provide improved forage for grazing cattle. Our study site would no longer be *shrub*steppe.

We thought of moving on to our next site. But it was late in the day, and we were hungry and thirsty. So we went ahead with our survey. We would no longer be able to include the site in our broader comparisons; now we would think of the range improvement as an unplanned experimental habitat manipulation.

Because we had previously surveyed birds and habitat at Guano Valley, we had a good baseline for a before-after comparison. That summer, the sagebrush was more robust but the vegetation otherwise had changed little. Abundances of sage sparrows, sage thrashers, and horned larks were essentially unchanged from previous years.[2] Numbers of Brewer's sparrows, which had varied considerably over the previous years, had decreased. That fall the sagebrush died. It was torn out and carried away and crested wheatgrass was planted.

We continued our surveys at Guano Valley for another three years. When we returned in the summer of 1981 the site was a barren wasteland. Only a short stubble of dead sagebrush remained, and the wheatgrass had yet to become established. Nonetheless, the birds were there, singing from the few remaining snags and defending territories. There were no changes in sage sparrow and sage thrasher numbers, although Brewer's sparrows continued their decline from the year before. A year later, however, Brewer's sparrows were again abundant, while sage sparrows began to decline. Three years after the treatment, sage sparrow numbers were even lower, while Brewer's sparrows were again scarce. What was going on?

Over the course of our surveys at Guano Valley, Brewer's sparrow numbers varied dramatically, as they did at other undisturbed sites in the area. Sage sparrows eventually responded to the disappearance of their habitat, but only after a time lag of two to three years. We attributed this time lag to site tenacity—the inclination of birds to return to territories where they had previously bred successfully. We knew from our long-term studies of marked individuals at Cabin Lake that sage sparrows may return to their previous territory over several successive years.[3] The persistence of distinctive song patterns in birds occupying the same territory locations at Guano Valley before and after the shrub treatment also suggested strong site tenacity. Apparently, some sage sparrows returned to occupy their territories of previous years even though their habitat (sagebrush) had disappeared, and they continued to do so for several years.

This "experiment," while unplanned, provided one more piece of the puzzle to help explain why shrubsteppe birds didn't seem to meet the expectations of the prevailing paradigm. In a variable environment, time lags in individual responses to habitat changes could erode close relationships to habitat features or the occurrence of other species. The expected patterns would not appear.

⁓

I noted before that reproductive success differed between the two sparrow species and varied among years. Many avian ecologists have proposed that such variations are associated with variations in the availability of food to feed young, which might be expected if food is a limiting resource. There was therefore a possible link with the resource-limitation thinking that had bedeviled our bird-community studies. John and I began to think of a way to explore this linkage experimentally. We couldn't alter the food supply for the birds in any realistic way, but we could manipulate the demand that they placed upon the resource. So one summer we conducted our own experimental manipulation.

Nesting Brewer's sparrows were abundant in the sagebrush at Cabin Lake. John and I asked what would happen if we increased the number of chicks in nests from the normal three to four, five, or even six chicks? Would the parent birds be able to keep up with the increased number of gaping mouths to feed? Would they be able to find enough food to feed their enlarged brood, or would food become limiting, setting the stage for competition?

We found nests in which the eggs hatched on the same day and transferred chicks from one nest to another to increase brood size. We then recorded the nest visits by male and female sparrows, what food they were bringing young, and how far they flew from the nest to forage for prey. This continued either until the chicks fledged or the nesting attempt failed. We weighed the young every day to compare their growth rates with chicks in unmanipulated control nests.

The results were clear-cut. Chicks in manipulated nests grew just as rapidly as those in control (three-chick) nests and fledged just as successfully, even when the brood size was doubled. The adults were clearly able to find sufficient food to meet the demands of their larger family. But it took considerable effort. The parent birds took shorter flights to foraging areas, spent less time foraging before returning to the nest, and generally brought back smaller prey items—they were less selective.

These results led to an obvious question: If Brewer's sparrows were capable of rearing more than three chicks in a brood, why didn't they? There were several possible answers. Perhaps clutch size was a fixed trait in the species (which, if so, simply deferred the question to another: Why should it be fixed at three?). Perhaps clutch size was determined by those occasional crunch years and could not be adjusted to times of plenty. Perhaps food was superabundant during the one year in which we conducted our experiments.

Our explanation was different. We had noticed that when brood sizes were increased, particularly to five or six chicks, the chicks nearly spilled out of the nest by the time they were ready to fledge. It seemed that the size of the nest cup was ideally sized to contain three young chicks ready to fledge. But in that case, why not build a bigger nest? Here, the short frost-free period at Cabin Lake entered in. When it was exceptionally cold (which happened for several days almost every summer), the female sheltered the young by sitting tightly over them with her body covering the nest cup. The nest diameter was a close fit to the body size of an adult Brewer's sparrow. A nest with a larger cup would leave gaps around the parent's body, increasing exposure of the chicks and the likelihood of nesting failure during cold spells.

The low level of predation at Cabin Lake and the apparent capacity of Brewer's sparrows to rear more chicks puzzled us. We'd sit around the campfire in the evening and, after a couple of beers, John would pose the question, "Why aren't we up to our armpits in Brewer's sparrows?" (He used a more colorful anatomical reference). If Brewer's sparrows could produce so many young, why weren't we seeing more birds? Over the years at Cabin Lake, we had banded several hundred sparrow chicks but had never recorded any returning to our study plots. Perhaps this shouldn't have been too surprising. Adults that we captured and banded did return to their territories for several years, so there may have been few openings for a returning chick from the previous year to wedge in among the established territories on our plots. Suitable sagebrush habitat extended for kilometers, however. If a returning young bird established a territory even a short distance from our plots, we would not have been likely to find it. Alternatively, perhaps mortality on the wintering grounds was high, so few of the young were available to return. Our question remained unanswered.

EPISODE: BIRDS, ARTHROPODS, AND PLANT CHEMISTRY

The experiments I've just described were attempts to shed light on some of the underlying assumptions of the competition paradigm—to determine *why* our findings in the shrubsteppe (and grasslands) didn't seem to fit. They were piggybacked onto our existing research, drawing on existing NSF support. Another field experiment, however, would take us in a different direction, for which we would need new dedicated grant support. We would need to convince our scientific peers that the experiment was worth doing.

The experiment was prompted by our observations of the use of sagebrush patches by foraging birds at Cabin Lake and our finding that larger patches contained more bird food. One fall, I gave an informal brown-bag lunch talk to my colleagues at UNM on what we'd been doing in the summers out in Oregon. I mentioned that we'd found that sage sparrows preferred to forage in larger sagebrush shrubs. I suggested that the birds might find more food there, not just because the bigger shrubs supported more insects but also because the shrubs might differ in some other respects.

After we'd finished lunch, one of the faculty, Rex Cates, pulled me aside.

"You know," he said, "sagebrush has a lot of plant secondary chemistry. A lot of that is thought to be defenses against arthropods. Maybe chemical differences among sagebrush shrubs are related to differences in the arthropods on the shrubs, to which the birds respond in their foraging."

"Let's talk," I said.

So we did. And as we talked, the idea of a collaboration began to form. Rex was a botanist whose focus was on plant secondary chemistry. His lab was equipped to analyze secondary chemicals present in plant tissues and he had technicians available to do the work. So we designed a field experiment to explore the linkages between plant chemistry, arthropods, and birds and fired off a proposal to NSF. Apparently, it was convincing; our proposal was funded, and we were on our way back to the Oregon shrubsteppe.

Rex and I recruited a field crew to spend a month or two during the summer living at our camp at Cabin Lake. Using individual shrubs as sample units, we designed the experiment to address several questions: Does the secondary chemistry of leaf tissues of individual sagebrush shrubs influence which arthropods occur on those shrubs?; Do arthropods, in a turnabout, influence the secondary chemistry of the shrubs?; Does bird predation affect the arthropod fauna present on a shrub, and thereby alter the secondary chemistry of the shrubs? These are the sorts of interactions that are at the core of ecology.

In an area several kilometers away from our long-term bird-study plots we tagged individual sagebrush shrubs and assigned them to one of three groups. One group included plants from which all arthropods would be removed ("defaunated") by enclosing the shrub for a few hours in a large plastic bag with pellets of aluminum phosphide. The pellets volatized quickly, killing all arthropods on the shrub. Half of the defaunated shrubs were then included in a second group that were enclosed in a cage of one-inch mesh crop protection netting. This kept birds away while allowing arthropods from nearby unaltered shrubs to move freely to colonize the shrubs. A third group of shrubs that served as controls were neither defaunated nor caged. We left the shrubs undisturbed after treatment for an interval of two to fifty-six weeks (the duration of the experiment before we called it quits) to see which arthropods colonized a shrub, when they did so, whether exposure to bird predation made a difference, and how any of this might be related to the shrub's secondary chemistry.

What did all the sampling, sorting, and identification of arthropods, and analysis of secondary chemistry tell us?[4] Arthropod groups differed in how rapidly they colonized sagebrush shrubs following defaunation. Several sapsucking taxa (e.g., hemipterans, homopterans) returned to control levels within two weeks of the defaunation. Others, particularly insect parasitoids and predators, recovered only after six weeks. Lepidopteran larvae and other leaf-chewing herbivores were slow to colonize the defaunated shrubs and remained scarce even fifty-six weeks after the treatment.

The changes in arthropods were accompanied by changes in secondary chemistry of the defaunated shrubs. By four weeks after the defaunation, several hydrocarbons, sesquiterpenes, and monoterpenes that occurred in most of the control shrubs were sharply reduced in the treated shrubs, while some alcohols and ketones that occurred at a low frequency in the controls increased dramatically after arthropods were removed. These changes persisted for the duration of the experiment. The changes in sagebrush chemistry on the defaunated shrubs were most strongly associated with changes in the abundance of herbivorous lepidopteran larvae, which fed on leaves and therefore were directly exposed to the secondary compounds in leaf tissues. Insects that fed on sap or phloem were not exposed to the secondary compounds and recovered quickly.

Rex and I pondered these results and (being good ecologists) came up with an interpretation: leaf-chewing herbivores, particularly lepidopteran larvae, directly affect foliage and may stress the plants. When defaunation removed these arthropods, the shrubs responded by shifting their allocation of secondary chemistry production to different compounds. And because recolonization of the defaunated plants by the herbivores was very slow, the changes in plant chemistry persisted.

So we found a strong, reciprocal linkage between arthropods and the secondary chemistry of sagebrush. What about the birds? In general, few arthropod taxa or functional groups differed between caged and uncaged sagebrush following defaunation. Counterintuitively, taxa known to be bird prey initially recovered from defaunation more slowly on the caged than the uncaged shrubs. But by the end of the experiment the diversity (but not the abundance) of bird-prey taxa was greater on the plants that had been protected from birds. The strongest response to caging, however, was by some fungi-eating arthropods (primarily mites), whose abundance doubled on caged versus uncaged plants by the end of the experiment. As the birds did not eat fungivorous insects, we were at a loss to explain this result. We found no differences in secondary chemistry between caged and uncaged shrubs.

The results of our experiment told us more about the relationship between arthropods and sagebrush secondary chemistry than about the role of birds as predators on arthropods. In a way, the caging experiment confirmed our earlier suggestion that bird populations in this shrubsteppe ecosystem were not limited by food availability, at least during our studies. If insect prey on sagebrush were limiting, then we should expect the plants protected by caging to develop greater numbers of arthropods by the end of the experiment. Birds didn't seem to have a strong impact on the arthropods and, consequently, the changes we saw in plant chemistry were independent of the caging.

The results of this experiment revealed something of the complexity of interactions among elements of shrubsteppe ecosystems. To pursue this further would require a deeper dive into the world of arthropods and plant chemistry. It was unclear what a follow-up experiment might look like, much less whether we could obtain additional funding. By the time we had written up our results for publication, both Rex and I had moved to different universities and further collaboration would have been difficult. And most of the work would be in the lab, not in the field. Following this pathway did not seem to be realistic.

What lessons can we take away from my years of doing ecology in the shrubsteppe? My studies had followed a progression typical of ecological investigations. I began by observing—getting a feel for the system and its birds. The observations then became increasingly quantitative, and the analyses plunged deeper and deeper into statistics. The results led to questions that prompted field experiments.

Conducting experiments out in nature, however, is not like experimenting in the lab. Even though you manipulate a clearly defined variable, the rest of the system is left free to vary. And it will. This doesn't doom field experiments, but it does mean that small or subtle responses to the experimental manipulation are unlikely to show through. Our manipulations were dramatic—doubling brood size in Brewer's sparrows or removing all the arthropods from individual sagebrush. Even then, the results required some creative explanations.

Field experiments are likely to be informative only if they are based on prior studies of a system. You can't conduct an experiment unless you have enough understanding of a system to hazard some predictions. Alternatively, you can conduct a brute-force experiment, altering the system to see what happens. This is what BLM did in Guano Valley, or what we did by giving Brewer's sparrows more mouths to feed.

In the end, I became disillusioned about the value of field experiments, at least in the shrubsteppe. Time lags in responses to the manipulations and interactions among multiple factors that varied in time and space made it difficult to design an experiment that would produce unambiguous results. This left us scrambling to come up with imaginative explanations. Although our experiments yielded insights, I doubted that they justified the massive efforts required to carry them out. I decided that this was, after all, not a pathway I wanted to follow. My interest was fading.

One of the most distinctive features of the shrubsteppe, however, is its spatial heterogeneity and patchiness. Our observations and experiments had taken advantage of the distribution of sagebrush as discretely separate patches. This theme of spatial patterning and patchiness was to lead me onto a new pathway, one that would dominate my activities and professional life for some time to come. It would lead me to become a landscape ecologist.

16

BECOMING A
LANDSCAPE ECOLOGIST

A
s my studies in the shrubsteppe wound down in the early 1980s, another pathway opened. I had been interested in environmental variability and its effects on birds for years. The temporal side of variability was expressed in the ways that weather and climate came to dominate my thinking about bird-community dynamics in grassland and shrubsteppe. The spatial aspects of variability first emerged when I considered heterogeneity as an important attribute of grassland habitats in my Wisconsin study. Heterogeneity continued to be an organizing theme as my research expanded into other grasslands and then into the shrubsteppe, where the clumps of sagebrush sharpened my focus on patchiness. Now I adopted these perspectives as the defining characteristics of a new pathway.

"Heterogeneity" is a rather nebulous term. In the broad sense, it simply means "not all the same." Heterogeneity takes on a more specific form when it is expressed as patchiness, which implies that variation is discontinuous and discrete. Patches are often internally homogeneous compared to the matrix in which they are embedded. Sagebrush shrubs in a matrix of bare ground or grasses are textbook examples. On a broader scale, remnant woodland or forest fragments in an agricultural landscape can be thought of as patches. My early emphasis on heterogeneity and patches was at odds with the competition paradigm in the 1960s and 1970s. Robert MacArthur had suggested that communities could best be studied in small, homogeneous areas where spatial variation did not confound the observations, and homogeneity had become closely tied to the presumption that ecological systems tended toward equilibrium. That hadn't worked for me; I saw heterogeneity and variation everywhere, at whatever scale I looked. This spatial and temporal variation bolstered my uneasiness with the paradigm. I was concerned that many ecologists did not seem to acknowledge the importance of variation, heterogeneity, and patchiness, so I began to write papers that developed the concepts and synthesized the ways in which they might affect ecological thinking. Looking back, I realize that I may have been reacting to the charge that our work in grasslands and shrubsteppe could be dismissed because it produced only anomalies. I also wrote these papers because I simply enjoyed thinking about concepts, probably even more than dealing with data and results.

I first addressed this issue of patchiness and variability in "Population Responses to Patchy Environments," a paper published when our work in the shrubsteppe was just getting underway.[1] Several of the themes I developed drew on my previous research. For example, I suggested that patchiness is the foundation for habitat selection: tadpoles in my experiments responded to square-patterned or stripe-patterned "habitat" patches. Habitat selection in a population may also be density-dependent: grasshopper and savannah sparrows in the Wisconsin grassland selected different habitat patches early in the season when population densities were low, but differences between the species' habitat choice disappeared later in the season as numbers increased. My discussion of predator-prey interactions emphasized how patch use by foraging individuals could be related to patch profitability and "giving-up time"—how long a forager should remain in a patch before moving elsewhere. This was a harbinger of my later observations of foraging by sage and Brewer's sparrows in sagebrush patches.

My interests in heterogeneity and patchiness predisposed me to follow a pathway into landscape ecology. After all, a landscape can be thought of as a mosaic of patches of different types, rather like pieces of a jigsaw puzzle. But in the 1970s, landscape ecology was not yet an identifiable pathway for me to follow. Although the discipline was becoming established in parts of Europe, only a few ecologists in North America were using the word "landscape" in combination with "ecology." Many of us interested in heterogeneity and patchiness were studying ecological relationships in natural (or at least seminatural) areas and may have associated the term "landscape" with the heavily modified agricultural or urban settings studied by European landscape ecologists. Whatever the reason, it was just a matter of time until the movement would take hold in North America. That happened in 1983.

EPISODE: THE OPENING OF THE LANDSCAPE ECOLOGY PATHWAY

It is rare for a scientific discipline to have an identifiable time and place of its birth. For landscape ecology in North America, however, the beginning was in a workshop, *Landscape Ecology: Directions and Approaches*, held at Allerton Park, Illinois, in April 1983.[2]

The workshop was organized by Paul Risser and his colleagues. Paul and I had been graduate students together at Wisconsin. We took several classes together, played basketball on the same intramural team, and shared ecological thoughts as we were "growing up." We had kept in touch. Knowing of my interests in spatial patterns and heterogeneity, he invited me to participate in the workshop. I probably would have gravitated to landscape ecology eventually, but his invitation was the tipping point that set me on that pathway. As I said earlier, people play a large role in throwing switches on the tracks of pathways.

There were twenty-five people at the conference, nearly all ecologists, many of them ecosystem ecologists with IBP in their backgrounds, and all (as an embarrassing sign of the times) white males. They were almost all in their mid-thirties or forties, old enough to be in mid-career, and open to the possibilities of a different approach to ecology. There was a strong feeling among the participants that thinking about landscape ecology had reached a "conceptual bottleneck," relying on observations and theories from other areas of ecology and geography with little of its own to distinguish itself conceptually. Accordingly, we spent three days (and evenings, over beers) hashing over ideas that might be fashioned into guiding concepts for the emerging discipline. It was great fun.

The themes that emerged set the agenda for landscape ecology in North America for the next several decades. Foremost was the emphasis on spatial heterogeneity—how it was generated; how it affected the flows of materials, species, or disturbances through a landscape; how it differed with scale; and how it might influence resource management by people. Landscape ecology was too new to have established a paradigm (it still doesn't have one), so the field was wide-open to exploratory research and new ideas. I realized that I was, in fact, a landscape ecologist; the emerging themes of landscape ecology were what I'd been thinking about for years.

Discussions at the workshop were sometimes intense. I remember arguing vehemently that *redistribution* should be recognized as a unique attribute of landscapes, something that could not be studied or understood without considering the interactions among patches in a landscape mosaic. Redistribution, as I envisioned it, was the process by which the abundance of materials, energy, or organisms in patches changed due to the movement of these things among patches. A focus on redistribution and how things moved would draw attention to the effects of patch boundaries and their permeability to movements and to the resulting changes in patch composition and dynamics. No one argued that these features of landscapes were unimportant; they just didn't consider them central to thinking about landscapes, as I did.

The discussions at Allerton Park defined the focus of landscape ecology as it became established in North America. Although the description and analysis of the spatial *patterns* of landscapes were essential components of landscape ecology, the emphasis was on landscape dynamics and landscape *processes* such as redistribution, movements, and the effects of patch boundaries. This perspective developed as a consequence of the people who gathered at Allerton Park, who were interested in ecosystems, vegetation dynamics, and population processes. A different group of ecologists might well have come up with a different blueprint to chart the course for landscape ecology in North America. There was a strong founder-effect on the direction of the discipline and, consequently, on the character of the pathway I would follow as a landscape ecologist. My own interests had undergone an evolution from an early focus on spatial variation (i.e., heterogeneity) to more explicit investigations of the effects of the patch-mosaic structure of shrubsteppe landscapes. Now I was poised to study how the patterns of landscapes affected ecological processes, especially as they were mediated by the behavior of individuals: their movements, responses to patch boundaries, and use of patches in a landscape mosaic—behavioral ecology. I was still an ecologist, but with a different modifier: "landscape" instead of "community."

Changing your disciplinary identification has consequences. You now have a different group of peers who will review your papers and proposals. You will attend different conferences. Your research proposals will be considered by a different program and your publications will be submitted to different journals. If you're in academia, your teaching responsibilities may change and you will attract different students for graduate training. These changes can be disruptive, so you shouldn't change your disciplinary identity without considering the consequences. And being sure that your interests are guiding you to do so.

I still had support from NSF to continue our studies in the shrub-steppe for another year. I realized, however, that moving down the new pathway would require separate funding. I roped two of my UNM colleagues, Cliff Crawford and Jim Gosz, into discussing what a new research program might look like. We immediately focused on patch boundaries: what factors cause them, how do they influence within- and between-patch processes, and how do the adjacent patches influence what happens at a boundary. These questions were the foundation of a proposal we submitted to NSF several months later.

We proposed an ambitious research program centered on boundaries that were defined by discontinuities in soils and vegetation in the semi-arid desert of the Sevilleta National Wildlife Refuge located south of Albuquerque. The three of us had complementary expertise: Cliff worked with invertebrates and decomposer food webs in deserts; Jim was an ecosystem ecologist whose research emphasized soil-plant-nutrient relationships; and I was (now) the landscape ecologist. We would sample invertebrates and soil chemistry across boundaries, conduct experimental manipulations of boundary sharpness and gradients, and supplement these studies with simulation modeling of movements of materials and animals across boundaries. Nothing in the proposal dealt with birds.

Despite the apparent irresistibility of a research plan relying on the complementary skills of three well-established investigators, the proposal was not funded. It was, in retrospect, a rather ponderous and (perhaps) overly ambitious proposal. Responding to the "conceptual bottleneck" in landscape ecology highlighted in the Allerton Park conference, we spent the first fifteen pages of a thirty-two-page, single-spaced proposal developing a conceptual framework for dealing with boundary dynamics in landscape ecosystems, emphasizing the themes of movements, redistribution, and boundaries. This may have been too much conceptual musing for reviewers looking for an explicit research design. The main criticism, however, seemed to be that the proposal was firmly grounded in landscape ecology, which was not yet established as a *bona fide* scientific discipline (at least in North America). There seemed to be a perception among the reviewers that it was closely allied with landscaping and urban planning. Moreover, I had labored long and hard to establish my credibility as an avian ecologist, but I was now proposing to do something different, something in which I had no prior track record. A different set of scientific peers was now reviewing my qualifications.

Undaunted by the rejection, we refashioned the proposal and submitted a scaled-down version a few months later. This too was rejected, for much the same reasons. In the interim, however, the three of us had converted the conceptual part of the first proposal into a manuscript in which we developed a foundation for landscape ecology based on the mechanistic details of individual movements and patch-boundary characteristics. The paper was in press by the time we submitted the second proposal, but that didn't seem to bolster the credibility of what we were proposing.[3] It was still heavily conceptual (but perhaps not mathematical enough to qualify as theory), and it was still landscape ecology. What we were proposing was not yet part of the scientific culture of ecology.

Cliff, Jim, and I continued to think about boundaries, but our quest for funding had been stymied. Cliff and Jim moved on to other research. In 1986, Bea and I left New Mexico to move to Colorado State University, where we busied ourselves establishing new graduate programs, teaching new classes, and initiating new research. I dabbled at formalizing some of the ideas about boundaries and patch dynamics, but I didn't get much further than that.

It might have remained so, but two years later Nils Chr. Stenseth visited the campus as part of a visiting ecologist speaker series that I organized. Nils Chr. was a population ecologist and theoretician from the University of Oslo in Norway whose primary research was with small mammal demography—also Bea's primary interest. Over several days we talked ecology together, Bea describing her studies with ground squirrels and me outlining my thinking about landscapes, movements, and boundaries. Our discussions must have piqued Nils Chr.'s interest, for a few weeks after he returned to Oslo we got a message: Could the two of us come to Oslo for a few months that autumn to continue our discussions? Once again, serendipity and encounters with key individuals would influence my opportunities to move along a pathway.

EPISODE: AUTUMN IN OSLO

It was an exciting offer. But both Bea and I had teaching obligations, and we weren't due for sabbaticals for several years. Nils Chr. could provide the funding to cover our travel and living expenses while in Oslo, but not our salaries. It didn't look promising.

We got together with our department chair, Bruce Wunder, to discuss possibilities, expecting him to list all the bureaucratic reasons this couldn't be done. But Bruce saw his role as making things happen to benefit both his faculty and the university. Spending time with Stenseth and his group would enhance our teaching and broaden our research. He suggested that we might shift our fall-term classes to the spring and double up our teaching. Then, rather than taking an official leave, we just wouldn't be around during the fall if someone came looking for us. It would be an *ad hoc* arrangement. We lined up one of our graduate students to stay in our house and take care of our dogs and we were off to Oslo with our three-year-old daughter, Kyra.

Because we were only going to be in Oslo for a few months, we stayed in a flat in graduate student housing on the edge of the city, where we could take a tram to the university and city center. It was delightful. It was only a short distance to a lake, Sognsvann, from which walking tracks took off into the adjacent forests. On weekends, the tracks were heavily used by Osloites, many pushing prams as they headed out for afternoon sweets at a café several kilometers into the forest. As the days grew shorter (and then very short), the afternoon light slanting through the trees was magical. We frightened and then delighted Kyra with tales of trolls. We found nests of wood ants that were taller than she was. And we learned to enjoy pickled herring and brunost (Norwegian goat cheese), but not gamalost, a pungent cheese that is best described as having the odor of dirty socks—definitely an acquired taste. We took the railway to the tiny mountain village of Finse (where Robert Falcon Scott had trained for the ill-fated *Terra Nova* expedition to the South Pole) to visit the Alpine Research Center and spent several days exploring the alpine tundra habitats. Trolls lurked everywhere.

Kyra joined a kindergarten/daycare, where she learned more Norwegian than we did. We spent much of the time at the university with Nils Chr. and his group. For me, it was an opportunity to revisit my conceptualization of boundaries and landscapes, now adding the perspectives on population processes that Bea, Nils Chr., and one of his students, Rolf Anker Ims, could provide. The result was a set of functional equations that set out the demographic consequences of movements of individuals within landscape patches; emigration of individuals across a boundary to other patches; and the patch-specific loss of individuals to predation and the like as they moved through a landscape mosaic. Our intent was to link landscape dynamics with ecological mechanisms—putting the "ecology" into landscape ecology.[4]

My interactions with Rolf went beyond brainstorming about populations and landscapes. Rolf was completing his Dr. Philos. degree while we were in Oslo and asked that I serve on his final examination committee. Of course I agreed, only to find out that at Norwegian universities the external examiners are officially "opponents," charged with aggressively challenging candidates (verbally) to defend their work and themselves. I took my role seriously, forcing Rolf to defend statements that usually might pass by unchallenged. It was not something I was comfortable doing, the more so because Rolf's parents were seated in the front row, unsmiling as I attacked their son. I tried to make him squirm, but Rolf defended himself and his work splendidly and passed the examination with distinction.

At the celebration that evening, there were many glasses raised in many toasts, generating much laughter. All in Norwegian. I didn't understand the jokes but felt relieved at no longer being considered an opponent. Rolf and I have remained good friends.

Landscape ecology provided a disciplinary home for topics that had interested me for some time. The ideas about boundary dynamics I developed with Cliff and Jim, and then with Nils Chr., Bea, and Rolf, built on my earlier interests in heterogeneity, patch use, and behavioral ecology. This thinking would become part of the conceptual foundation of landscape ecology, part of its developing scientific culture.

17

DEALING WITH SCALE

The importance of boundaries and redistribution were not the only themes emerging from the Allerton Park conference that meshed with my interests and influenced my thinking. The discussions at Allerton also touched on how to incorporate the findings and insights from the m² spatial scales at which ecological studies were usually conducted into understanding the patterns and processes displayed at the broader, km-wide spatial scales of landscapes. *Scale* was emerging as a central focus of landscape ecology.

Scale had been attracting increasing attention among ecologists as challenges to the competition/equilibrium paradigm mounted. When the paradigm held sway during the 1960s and 1970s, many ecologists were content to assume that habitats were fully occupied, in which case the results of ecological studies conducted in a small homogeneous area could be extrapolated to areas of like habitat at broader spatial scales. A parallel assumption applied to time—if systems were normally at or close to equilibrium, results from a short-term "snapshot" study should apply to other times or longer periods as well.

Those were the assumptions. But if environments and communities were not in equilibrium and habitats were patchy, then scale could not simply be ignored. The results of a study in one year at one place might be quite different in another year at the same place or at a different place in the same year. Patterns evident in a small area might disappear or change into something else when considered over a larger area. In variable systems, scale matters.

This should sound familiar—it's what we found repeatedly in our studies of birds in grasslands and shrubsteppe. As the abundances of species in study plots varied from year to year, comparisons among plots could produce strikingly different patterns of bird communities depending on which years and plots were included in the comparisons. Habitat associations of species varied depending on the scale of analysis. The relationship of sage sparrow abundance to sagebrush, grass, or bare-ground coverage, for example, differed when considered over a broad biogeographic scale versus a regional or local scale.[1] What we saw looking through the window of our analysis depended on the size of the window—its scale. Unless these inconsistencies in habitat relationships were manifestations of the weirdness of the shrubsteppe, they implied that the results of investigations might depend on the scale of study. Consequently, extrapolations from one scale to another could produce misleading or incorrect results. Some of the contentious arguments that preoccupied population and community ecologists from the 1960s into the 1980s probably arose because they were studying similar things at different scales. When their results did not match, ecologists argued over who was right rather than considering whether the differences might be due to differences in the scales of their studies.

My interest in spatial and temporal variation and heterogeneity predisposed me to wrestle with the problems posed by differences in the scale of investigations. I had offered some initial musings about scale when I reviewed population responses to patchy environments in 1976, and I later expanded on these thoughts when I considered how differences in scale might affect the interpretation of censuses of bird populations.[2] The issue continued to nag me (and other ecologists), but it wasn't until years later that I was able to focus my thinking on scale, and then only because of an unusual opportunity. Serendipity struck again!

EPISODE: SCALING IN THE ROCKIES

In the summer of 1987, I was invited to deliver the inaugural Katherine P. Douglass lecture at the Rocky Mountain Biological Laboratory (RMBL). The laboratory is at 9,000-plus feet (2,750 m) elevation at the site of the ghost town of Gothic, a remnant of a short burst of silver mining in the Colorado mountains during the nineteenth century. In the summer, RMBL is a hotbed of high-altitude ecological research, attracting scientists from throughout the world. The lectureship brought me to RMBL for a week to hobnob with the scientists and students and give an evening lecture on whatever I wanted to talk about. I decided this was a chance to see if I could fashion some sort of conceptual framework for dealing with scale that might be useful in ecology and, especially, landscape ecology.

RMBL was an idyllic setting for thinking. The laboratory was surrounded by the peaks of the West Elk Mountains and the alpine meadows were ablaze with wildflowers. One of my graduate students, Natasha Kotliar, was studying how hummingbirds used the patches of flowers in their foraging, so we spent several days at her field sites, watching the birds, gathering data, and talking about heterogeneity, landscapes, and scale.[3] Meals in the dining hall provided opportunities to bounce ideas off a diverse group of ecologists of all ages, and in the evenings I could retire to the rustic cabin I had been given, to consolidate my thoughts. When the time came to give my lecture, I felt I was ready.

I emphasized three challenges that scale creates for ecological investigations. First, the appropriate scale for a study depends on what questions are being asked; there is no single "right" scale. Natasha could study the use of flowers by foraging hummingbirds at a scale corresponding to the size of individual territories; an investigation of the factors influencing hummingbird population dynamics would require a quite different scale.

Second, different organisms perceive and respond to the environment at different scales. It's trite to point out that elephants and mice relate to the same environment at different scales, requiring that studies of each be scaled accordingly. But do the meadowlarks and Brewer's sparrows we studied, which differ in body size by an order of magnitude, scale their use of habitats in the same way? Observations should be made at scales that are appropriate to the organisms and phenomena being studied. If multiple species are being considered (as in a community or ecosystem study), should you use a single arbitrary scale and risk coming up with incorrect results because the scale doesn't match that of some species, or should you adjust the sampling scale to the scales of the organisms and risk producing incompatible data because the observations were gathered at different scales?

And third, ecology at its core is concerned with determining what processes produce the patterns of nature: process-pattern (or cause-effect) relationships. These relationships may be scale-dependent, but not in an orderly way. I suggested that we might think of "domains" of scale within which the linkage between process and pattern remains constant or changes in a continuous (linear) way. As scale broadens, however, there may come a point at which other factors intercede and the process-pattern linkage changes rather abruptly (nonlinearly). For example, plant growth might be determined by soil-nutrient levels in a similar way over a range of fine spatial scales, but at some point, broader-scale factors such as slope or exposure might override those effects. Generalizations about ecological processes and patterns could be made over scales within a domain, but extrapolations across a threshold between domains would be problematic because of the mixing of different process-pattern relationships.

These ideas about scale would all make ecological investigations more difficult. However, scale could not easily be ignored; it would need to be considered in the design of any ecological project. My lecture generated considerable discussion, as was my intent. One response, however, was unanticipated. Paul and Anne Ehrlich spent their summers at RMBL, and Paul was in the audience. He quickly jumped in.

"You haven't said anything new," he said. "Ecologists have known all that stuff for years."

Perhaps I should have had a snappy response, but Paul was (and still is) an iconic figure among ecologists and evolutionary biologists. I probably mumbled (feebly) something like, "That's not what I see when I look at the literature."

Paul and I didn't continue the conversation, and I came away wondering whether I really had anything worth saying after having spent the week (and several years) thinking about scale. But Ward Watt came to my rescue. Ward was an evolutionary geneticist who, like Paul, was from Stanford. He consoled me, suggesting, "That was just Paul being Paul." Ward was also editor of the journal *Functional Ecology*, published by the British Ecological Society. He suggested that my talk (or some version of it) might make a nice paper for the journal and encouraged me to keep at it. So I did. A few months later I submitted the paper. It was published in 1989 and has become widely cited by both ecologists and landscape ecologists.[4] There must have been more to my ideas than Paul thought.

My recognition of the variability of the grasslands and shrubsteppe that I studied had drawn me to think about scale. The implications of scale extended beyond community ecology or the competition paradigm. If the findings of ecological investigations are scale-dependent, as I was proposing, then studies conducted at different scales could not simply be combined as if scale didn't matter. Ecologists would need to select the appropriate scale for their investigations and be attentive to the potential effects of scale differences when drawing comparisons among studies.

This was not what many ecologists wanted to hear. I was once again challenging well-established ecological practices, this time based on concepts and (I felt certain) logic rather than the empirical results of my studies. But I was not alone. Other ecologists were also calling attention to the confounding effects of scale. Before long, awareness of the importance of scale would lead to a shift in thinking among ecologists. Within a decade, symposia and conferences were being organized to consider scale and books about scale were being written. My pathway in landscape ecology would now include a consideration of scale as well as the mechanistic focus on patchiness and movement behavior that emerged from our collaborations in Oslo.

18

FOLLOWING THE LANDSCAPE ECOLOGY PATHWAY

Landscape ecology emphasizes how the patches in a landscape mosaic interact with one another, how the structure of a landscape influences a wide variety of ecological processes and patterns, and how it all depends on scale. The conceptual framework that Nils Chr., Bea, Rolf, and I developed in Oslo focused on how the movements of individuals might be affected by landscape structure. The movements of individuals, in turn, could lead to spatial variation in foraging behavior and prey encounter, mating opportunities, predation risk, interactions with competitors, the distribution and dynamics of populations, and many other ecological phenomena. I was thinking that movement might be the core process that integrates "landscape" with "ecology."

The first step would be to link movements to landscape structure. To do this, I would have to go beyond simple measures of habitat heterogeneity. I would need to relate the movements of individuals to patches, boundaries, and the other details of landscape structure, and to do this I would need to track in detail the locations of individuals as they moved through a landscape. Ideally, I would also manipulate landscape structure in specific ways to see how individuals would respond. I would conduct landscape experiments. The birds that I had been studying for most of my professional (and preprofessional) life were lousy subjects for such investigations. In the grassland and shrubsteppe I had charted how individual birds moved among patches in their territories, but I couldn't see how I could track their movements in detail, especially as they flew where they wanted throughout the landscape at a broader scale. Reluctantly, I realized that if I were to follow this pathway birds wouldn't do. I'd need to find a different system.

Our discussions with Rolf Ims while Bea and I were in Oslo had touched on the difficulty of conducting experimental manipulations while retaining some degree of control over critical variables at the scale of "real" landscapes. We had considered how "experimental model systems" (EMS), which mimic a complex system by reducing the scale and simplifying the processes, could be employed in landscape ecology. Perhaps I could realize the benefits of an EMS by scaling down the spatial pattern of landscapes using fine-scale landscapes—"microlandscapes"—that covered several m^2 rather than km^2. In this way, I could experimentally manipulate landscape structure, replicate observations, and take detailed measurements. Larger sample sizes would enhance statistical rigor, which would please the reviewers of my grant applications and research papers.

But what organisms would be likely to respond to landscape structure at this finer microlandscape scale? After considering several possibilities, I settled on beetles, specifically darkling (tenebrionid) beetles of the genus *Eleodes*. Why these particular beetles? Darkling beetles are generalist detritivores and herbivores that are abundant in the desert grasslands and shortgrass steppe where I intended to conduct my studies. They are conspicuously black, diurnal, large, and walk rather than fly. They can be followed and tracked as they move across a microlandscape of patches of vegetation and bare ground.

But beetles also have some disadvantages. When I watched birds, I generally had some idea of why they were doing what they were doing—attracting mates, searching for food, defending territories, feeding young, and so on. Birds are visual creatures, so I had some confidence that they saw the habitat pretty much as I did. My measures of habitat composition and structure would therefore be appropriate. I hadn't a clue how a beetle perceived its environment or which sensory modalities they used. Consequently, as I followed them and tracked their movements, I would have little idea why they were going wherever they were going. However, because I wanted to see how movement pathways, time spent in patches, and behavior at patch boundaries were affected by features of a landscape, I felt that I could ignore whatever factors might be motivating beetles as they wandered through what, to them, was a landscape. I was interested in pursuing landscape ecology from a beetle's perspective.[1]

At this time, I was still at UNM. I was trying to figure out how to fashion movements, landscapes, scale, and beetles into a coherent research program when Bruce Milne joined the Biology Department faculty. Bruce had been trained as a landscape ecologist. Although we overlapped only briefly before I took off to Colorado, it was long enough for us to think of how we might collaborate. That's when my pathway into landscape ecology took off in an unexpected direction.

Bruce was enamored with fractal geometry. Formal treatment of fractals goes well beyond what I can describe here.[2] Suffice it to say that fractal dimension can be used as a measure of the spatial complexity of a mosaic of patches in a landscape or the tortuosity of an individual's movement pathway as it traverses a landscape. Bruce and I decided to use the tools of fractal geometry to explore how animals move through a landscape, using beetles as an EMS. We submitted a collaborative research proposal to NSF, which was rejected. Reviewers apparently still had reservations about whether landscape ecology was really science. But the jazziness of fractals, our focus on scaling, and our intent to conduct designed experiments tipped the balance and our proposal was approved on the second try. The funding enabled us to support several graduate students and a postdoc, Alan Johnson.

Fractals, fine-scale EMS, and beetles became the linchpins of my research program in landscape ecology.

My pathway in landscape ecology shifted because of a new collaboration (the effects of *key people*) and the approval of our NSF proposal (the influence of the *scientific culture*). Had neither of these forces come into play it's unlikely that I would have headed in this direction. The messages are clear: seek out collaborations that interest you, and don't give up on your efforts to carry out the research to address the new questions.

EPISODE: BEETLES AND FRACTALS

By the time our collaborative research finally got underway, Bruce and I were at different universities, so we launched parallel programs at UNM and Colorado State. Bruce and Alan focused field efforts in desert grasslands at the Sevilleta National Wildlife Refuge located south of Albuquerque, while I initiated studies in the shortgrass prairie at the Central Plains Experimental Range (CPER) located northeast of Fort Collins (the same Pawnee site I had used in my IBP work). At each site we mapped the vegetation in a 25-m^2 plot (a beetle-relevant "landscape" scale) and calculated the fractal dimension of vegetation patchiness at multiple scales. We then released beetles individually in the center of the plot and charted their movement paths. We wanted to see how the landscape pattern affected how a beetle moved through the plot. Because beetles wander a bit even in a uniform area, we simulated their movements as a correlated random walk, in which the direction and features of one segment of a pathway are correlated with those of the previous segment—an individual moving in one direction is likely to continue moving in that direction. We could then determine how often simulated beetles would be likely to encounter areas of a given fractal dimension in the plot in comparison to our observations of real beetles, which would tell us how the fractal dimension of the microlandscape affected beetle movements.

What did all this measurement, analysis, simulation, and fractalizing show?[3] At Sevilleta, real beetles and simulated beetles encountered areas of relatively low and high fractal dimension (i.e., mostly grass cover or mostly bare ground, respectively) with similar frequencies, as if they were following a correlated random walk. However, areas with intermediate fractal dimensions, where grass clumps occurred as discrete patches in the matrix of bare ground, were visited significantly less often by real than by simulated beetles. Beetles moving over bare ground shifted their movements to avoid the boundary with grass patches and turned to continue moving over bare ground. And they did so even when the patchiness was expressed at different scales.

At CPER we expanded the study to include three *Eleodes* species that differed in body size and, correspondingly, the distance traveled during a time interval.[4] All three species responded to differences in landscape structure in similar ways, moving continuously and rapidly over the grass sward (lower fractal dimension) but following more tortuous routes (higher fractal dimension) where patches of other vegetation interceded. The movements of real beetles were more directed and the distance traveled per time interval was greater than in the correlated-random-walk simulations. As at Sevilleta, the fractal dimensions of beetle pathways were similar at all scales that we considered.

꩜

Following our initial investigations at Sevilleta and CPER, Bruce and I headed in different directions. Bruce and Alan continued to use simulation models and field experiments with beetles to show how theories from other disciplines and the methods of fractal analysis could contribute to understanding issues of scale and movements in landscape ecology. I was not as enthusiastic about fractals as Bruce and Alan. Fractals could express how the complexity of a movement pathway might relate to the complexity of a landscape pattern, but if one found a relationship, what did it mean?

Did fractals tell you anything about how beetles perceive and respond to their environment? Fractals could indicate whether the complexity of a pathway changed depending on the scale on which it was viewed, but then what? In other words, fractal geometry provided a tool to quantify the complexity of a pathway or a landscape pattern, but its ecological interpretation remained elusive.

Despite my reservations, my students continued to explore the nuances of fractal measures and use microlandscapes and beetles (or other arthropods) as EMS (see chapter 19). But I wasn't motivated to continue in that direction. My attention shifted from empirical investigations to thinking about how landscape ecology could contribute to or draw from other areas of ecology. I was at the stage in my career when scientists are often inclined to reflect on the broader implications of their research. I felt that I might be able to contribute more to the development of landscape ecology as a science by conceptualizing, synthesizing, and writing (also known as "pontificating") than by undertaking yet another empirical study.

EPISODE: SPRINGTIME IN DARWIN

I had an opportunity to expand my thinking about landscapes in 1996, when Bea and I were invited to visit Australia for a few months to learn more about how ecology played out in Australian landscapes. Alan Anderson asked if I could spend some time with colleagues at the CSIRO Division of Wildlife and Ecology and John Woinarski invited Bea to do the same at the Parks and Wildlife Commission of the Northern Territory. I would share my perspectives on landscape ecology and scale and Bea would focus on wildlife ecology and population dynamics. Both laboratories were on the same grounds in Darwin, so the logistics (once we got there) would be easy. It was an exciting prospect, but once again there was a complication: we weren't due for sabbatical leave for a few more years. And once again our department chair (now Joan Herbers) stepped in. She arranged for us to double up our teaching and just not be around for a few months while we were in Darwin. She made it possible for us to take advantage of the opportunity.

So we headed back to Australia, where it was the dry season (the "Dry") in the "Top End" of the Northern Territory. We moved into a cottage on the CSIRO grounds, enrolled our daughters (Kyra, now ten, and Taryn, six) in school, and signed them up with the local swim team. The ten-minute walk from our cottage to the labs passed through a mango orchard that had been used in horticultural experiments and was now abandoned, so we could pick a ripe mango or two for lunch on our way in the morning. There were goannas and frilled lizards about, and great bowerbirds, grey butcher-birds, and a variety of fairy wrens and honeyeaters frequented the shrubs around our cottage. A seasonal wetland that was great for walking and birding was a short distance away, and mangrove tidal flats and the ocean weren't much farther. However, one did need to watch out for saltwater crocodiles—"salties"—and swimming in the ocean wasn't a good idea when box jellyfish were about. Australia abounds in dangerous and venomous creatures.

Eucalypt savannas dominate the landscapes of the Northern Territory. Monsoonal rains during the November to April wet season (the "Wet") promote rapid growth of grasses in the understory, which become tinder-dry during the following Dry. Consequently, fire plays an overwhelming role in shaping the landscapes. Most areas burn once or several times a year, and fire is a major tool in landscape and habitat management.[5] Much of the research at the CSIRO lab centered on fire in the savannas. I talked with the researchers, read their papers, listened to their presentations, and visited their field sites. All of this gave me a real-world appreciation for the sometimes-abstract concept of patch dynamics. As the Dry progressed into the buildup to the Wet, lightning strikes increased in frequency and more small fires were ignited by managers or local stakeholders. The landscapes of the Northern Territory became a shifting mosaic of savanna vegetation in different stages of incineration.

In the forests of the western United States where I lived, we tended to think of fire as a disturbance whose frequency and intensity had been altered by decades of mismanagement. In contrast, the dynamics of fire and the spatial patterns they created in the Australian savannas were not recent developments. Aboriginal peoples had used fire to manage habitat and wildlife populations for millennia before Europeans entered the scene, long enough for human-caused fires to become a force shaping the evolution of savanna vegetation and life-history adaptations of many animals. The linkages between fire, people, climate, and savanna ecosystems were tight and ancient. There was a long-term equilibrium of sorts underlying the dramatic annual variations in climate, vegetation, and fire that were aided by human actions. Ecological dynamics in the savannas were relatively stable or highly variable depending on the time scale on which they were viewed.[6]

Bea and I took advantage of opportunities to spend time in the field with scientists studying wildlife populations (Bea) or fire ecology (me). We also wanted Kyra and Taryn to experience more of the Northern Territory than just school and swimming, so we took several excursions into the areas around Darwin. The most memorable was the week we spent in Arnhem Land. Access to most of Arnhem Land is restricted to its Aboriginal custodians, but dispensation had been granted to a small safari camp at Mount Borradaile. We flew there from Darwin in a lightplane, landing on a dirt strip carved out of the savanna woodlands. Aside from our Aboriginal guide, Sid, and a small staff, we were the only people there. We took walks through the bush and spent time at a billabong teeming with wildlife. It was one of the few places to still have water at the end of the Dry, so thousands of pied geese, whistling-ducks, and other birds had aggregated there. Sid assured us that there was more to be seen, so he took us out in a small boat one evening. After dark, he suggested we shine a spotlight on the shoreline. Hundreds of eyes reflected back at us—crocodiles. We decided not to go ashore there.

We spent most of our time at Mount Borradaile visiting the rock shelters around the camp. Aboriginals had occupied many of these shelters for thousands of years.[7] They were spectacular galleries of Aboriginal rock art—layers upon layers of intricate images. Some showed mythical figures such as Namarrgon (lightning man) or the Rainbow Serpent, others depicted extinct animals such as thylacines or important food sources such as barramundi. Some had been done within the past few decades, others (e.g., a ship or a person with a gun) recorded earlier encounters with other cultures, and some had been made more than 50,000 years ago. The abundance and diversity of art was an overwhelming reminder of the richness of Aboriginal culture and life. Landscapes in Australia have included and been shaped by people for a very long time.

As landscape ecology began to be recognized as an emerging discipline both nationally and internationally, I was invited to participate in conferences and symposia and contribute chapters to books. My pathway was still firmly centered on landscape ecology, but it was now expressed conceptually rather than empirically. My presentations and writings promoted landscape ecology as a distinctive approach to ecology, one involving spatial analysis, scale, and animal behavior and built on the foundation principle that the spatial structure of a landscape mosaic matters—it affects all manner of ecological patterns and processes. Just as ecologists' earlier assumption of spatial homogeneity had given way to concepts of heterogeneity, "heterogeneity" was now being refined by a consideration of the explicit spatial structure and configuration of landscapes at multiple scales.

At one point, for example, I was invited to a symposium in France dealing with dispersal and its evolutionary consequences. Most of the other participants were ecological geneticists or evolutionists, to whom dispersal was a key process affecting gene flow and the genetic distinctiveness of populations. Dispersal was visualized as the departure of individuals from point A (generally their birthplace) and the probability of their subsequent arrival at point B (generally where they reproduced). Many models treated this probability only as a decreasing function of the distance between points

A and B. I pointed out that this probability would also depend on what happens to individuals between points A and B, which would be influenced by the structure of the intervening landscape.[8] Rather than following an assumed straight-line pathway between A and B, dispersing individuals might follow a more tortuous route, responding to patches and boundaries in the landscape mosaic much like the beetles that we tracked in the grasslands. Variations in landscape structure could have profound effects on dispersal probabilities, and consequently on dispersal behavior and its evolutionary consequences. "Ignore landscape structure at your peril," I warned. The message was not one that evolutionary ecologists and geneticists wanted to hear.

This theme—of how landscape ecology could offer ways to deal with the reality of the multiscaled patchwork mosaic in which most organisms lived—came to dominate my writing. I wrote about such things as how landscape ecology could contribute to thinking about metapopulation dynamics, habitat fragmentation, conservation, resource management, and population dispersion patterns; how to incorporate scale into ecological experiments; and even how to apply concepts from landscape ecology to the study of riverine ecosystems. I also wrote more broadly about landscape ecology as a discipline: why it was relevant, how it might blend scientific and cultural perspectives on landscapes into a unified approach, and what I considered to be its central concepts. Not surprisingly, these concepts emphasized the aspects of landscapes that had interested me for decades: the importance of variations in patch quality, boundaries, the context of patches in a mosaic, connectivity, movements, and the scale-dependence of spatial patterns and processes. Someone else might have highlighted different concepts, particularly if they followed the European approach to landscape ecology. I admitted that consideration of these facets of landscape dynamics would make ecological investigations and conservation efforts more complex but concluded that the default position had to be to consider them important unless there were good reasons to think otherwise.[9] There was a bit of *déjà vu* here: when I was following a pathway in community ecology earlier in my career, I had called attention to how variation in space (and time) would complicate attempts to come up with satisfying generalizations. Now I was suggesting that landscape ecology might similarly complicate ecological and evolutionary investigations more broadly. Apparently, I carried my iconoclasm with me as I moved from one pathway to another.

Over a span of nearly two decades, my progression along the landscape ecology pathway took me from some initial conceptual musings stimulated by the Allerton Park conference to fractals and empirical observations and experiments with beetles in microlandscapes, and then to a flurry of writings promoting landscape ecology as a legitimate and important science. It even led to a deep involvement in the professional organization of the discipline.[10] Such a progression is not unusual for a scientist entering middle age (as I was). But there's always the potential for some of the things that interested and excited you and set you on a pathway to begin with to be abandoned or sidetracked the farther you progress along the pathway.

I had started on this pathway because I was excited about the way landscape ecology consolidated my interests and thinking about spatial variation and behavior. It offered a way to test some of my ideas using EMS. I would continue following that pathway, at least vicariously, through my students, even as my interests shifted elsewhere.

19

TRAVELING ALONG PATHWAYS WITH STUDENTS

Being a scientist—an ecologist—isn't just about following your interests and opportunities to conduct research, publish papers, and become part of a scientific culture. That's the research side of things. But if you're in academia (as I was), there will be classes to be taught and seemingly endless committee meetings to attend or administrative duties to fulfill. How your time will be weighted among these three functions—research, teaching, and service—will vary, but most academic positions entail all of them. And they blend into one another. The excitement of research carries over into teaching. Teaching, in turn, may generate ideas that lead to productive research. And service provides a way to infuse your thinking and experience into university programs and policies, which may enhance teaching and bolster research opportunities.

I've emphasized research in describing the forces and factors that determined the pathways I followed in becoming an ecologist. What about teaching and service? In many respects, the greatest joy of being a faculty member is working with students—teaching. You can excite students about things that excite you; share your knowledge, experiences, and perspectives; and, if all goes well, see them become, if not scientists themselves, at least scientifically literate.

Teaching can be both frustrating and fulfilling. You may lecture to a large class, believing that students are following your words with rapt attention. The reality is that some of them would rather be someplace else. They might attend your class for a variety of reasons: because it's a requirement of their degree program; it is a prerequisite for another class they really want to take; it fills a gap in their schedule; they've heard it is a snap course or has a charismatic teacher; or because they are actually interested in the subject. Here's the opportunity to kindle that interest. You can deluge them with information about things that are near and dear to you and expose them to a scientific way of evaluating evidence and arguments—of thinking. And occasionally you can help them find their own pathway (as people like Bob Burns, Chuck Carpenter, and George Sutton did for me).

It didn't happen often, but several times someone approached me at a meeting and said, "I bet you don't remember me."

I usually didn't. But they continued, saying something like, "I took your ornithology class when you first came to Oregon State, and it inspired me to get a degree in wildlife management."

"So what are you doing now?" I replied.

"Oh, I've been working on endangered species management in a state agency for the past thirty years. I'm now retired."

Aside from making you acutely aware of the passage of time, such moments make it all worthwhile.

～

My moves to join the faculties at the University of New Mexico and Colorado State University were prompted by their desire to strengthen their ecology curriculums by hiring a senior ecologist. This turned out to be me. I would do my part by conducting research (and bringing in research funding), attracting graduate students, and teaching graduate-level courses. But I could also work with other faculty and the university administration to enhance ecology on the campus—a service role. At UNM this involved developing several new course offerings that integrated ecological concepts and practices about populations, communities, and ecosystems. Because most ecologists were housed in the Biology Department, implementing new courses was administratively easy.

At Colorado State, the opportunities and the challenges were both greater. There were many ecologists on the faculty, but they were scattered among multiple departments.[1] "Ecology" was not a coherent strength of the university. The seeds of integrating ecology into a campus-wide program had already been planted, however. Indeed, that was one of the things that brought me to Colorado State. Now the challenge was to bring it all together.

Several of us decided that the best way to capitalize on the extraordinary strength of ecology on campus would be to offer graduate degrees—MS and PhD—in ecology. Graduate degrees are usually conferred by academic departments, but there was no Department of Ecology. Forming a large, interdisciplinary department to grant degrees was not what we wanted. That would only create more bureaucracy. Moreover, several department chairs saw our suggestion as a threat that would draw faculty, resources, and prospective students away from their programs. We needed to develop a mechanism for graduate training and granting degrees that fell outside the normal academic departmental structure.

Dealing with the reality of campus politics fell to me. I met with chairs and deans to explain what we had in mind, and then took their reactions back to the drawing board. We kept trying. It took several years, but we eventually created a program—the Graduate Degree Program in Ecology (GDPE)—that could grant graduate degrees while retaining the departmental structure for faculty affiliations and teaching.[2] The program provided a way to coordinate and integrate ecology research and teaching across the campus. Students could now create a program of study that would enable them to become ecologists rather than trying to fit ecology into the framework (and requirements) of a traditional departmental program.

The program would attract students who wanted to become ecologists.

⁓

At Colorado State, most of the graduate students working with me took their degree in GDPE—they wanted to become ecologists. At Oregon State and UNM, they were able to cobble together academic programs that accomplished the same goal, but without the ecology label. The students who applied to join my program came because it matched their interests, or sometimes because they were attracted to the school or location or were following a spouse or other attachment. Sometimes they had heard of my reputation and applied because of (or despite) that.

Over the years, the students who came to work with me had a wide variety of interests and goals, which was what made things so interesting and exciting. Regardless of their motivation, I aimed to immerse them in the world of ecology, to guide them into doing research (as John Emlen had done for me), and to provide a collegial and stimulating setting in which they could grow and prosper. I wanted to train them how to play the game of science and be a professional. I taught them the nuts and bolts of ecology, of course, but more importantly I wanted them to learn how to ask questions, figure out how to get answers, and know what to do with the answers when they had them—how to tell others about what they found and, in the process, sell their science and themselves.

The graduate students who joined my program studied a wide variety of organisms and topics. Most worked with birds, of course, but others studied bighorn sheep, chipmunks, voles, desert lizards, scorpions, grasshoppers, ants, cactus bugs, dung beetles, and even crabs. While many focused on some aspect of behavior (feeding, reproduction, mating systems, song structure), others addressed questions in biogeography, community dynamics, or species interactions. Several investigated how animals responded to landscape structure and the effects of scale, exploring what insights fractals could reveal. My role for all these students was to be a mentor—to be a sounding board for ideas, encourage them when things didn't go as planned, and help them find their own career pathways. Many took the academic route, going on to a postdoc or a position with a university. But that was not their only pathway. Some joined a state or federal resource management agency. Yet others followed a different path: one became an ecotour guide, another an environmental lawyer. Several entered environmental consulting, and one even became a college president.

The interactions were not always one-way, however. Sometimes their interests and choices opened potential pathways that I might also follow. My role then became as much collaborator as advisor. This brings me to ants.

Tom Crist joined our group as a postdoc in 1990. He had conducted his doctoral research on the spatial dynamics of foraging by harvester ants (*Pogonomyrmex*), so he immediately saw an opportunity to expand our work with beetles to include ants. Harvester ants are especially well suited to spatial analysis. Colonies occupy nests from which individuals fan out to collect seeds, insects, and other detritus. The ants clear vegetation from an area of a meter or two around nest mounds, making the colonies easily visible in aerial or satellite images. And *Pogonomyrmex* colonies are abundant in the grasslands at CPER.

So we added harvester ants to the menagerie of insects we were studying, charting their movement pathways, recording the spatial distribution of ant colonies, and assessing how landscape structure at several scales affected ant individuals, colonies, and populations. Initially we thought we would follow individual ants as they moved for an hour or so, much as we had done with beetles. Keeping track of individual ants among the hoards moving about a colony, however, would prove difficult. We came up with what we thought was a clever way of marking individuals so they could be followed. We caught an ant, put it in a vial with fluorescent powder, and shook it to cover the ant with powder. We then released the ant (now bright pink) and began to follow it. The marked ants quickly headed back to the nest mound and entered the underground colony. They just as quickly reappeared, carried out of the nest by other ants who deposited the marked ant at the edge of the cleared area. No matter how often the marked ant tried to return to the nest it was carried away. Apparently, the fluorescent powder interfered with the cues that ants use to recognize colony mates and the marked ant was treated as a bit of detritus to be cleaned from the nest. Our cleverness didn't fool the ants. We ended up following unmarked ants for shorter periods.

Fluorescent powder aside, we used the same methods to record movements of beetles, grasshoppers, and now ants. Although these organisms differed greatly in body size and thus in movement rate and path length, fractal dimension provided a common measure of path complexity. Faced with the same microlandscapes, the taxa differed in the complexity of their movement pathways: beetle pathways were generally more complex than those of grasshoppers, and ants followed even more tortuous routes.[3] These differences may have reflected differences in how the taxa used resources. When a foraging ant found a patch of seeds it would usually continue searching in a restricted area—a highly convoluted pathway (high fractal dimension). Beetles and grasshoppers fed on more widely scattered resources (detritus or vegetation) and followed more linear search pathways. In this case, fractal dimension provided a simple and interpretable measure of movements that might otherwise be obscured by the size-related differences in other pathway measures, but it also needed some natural history insights to tell you what it meant.

Tom's enthusiasm for ants was infectious. We conducted experiments to assess how vegetation patchiness affected seed harvesting by ants, analyzed the dispersion of ant colonies at multiple scales, and assessed how ant movement patterns could affect estimates of population sizes based on pitfall trapping. I was beginning to see why people like Ed Wilson found ants so fascinating. Perhaps ants would join birds and beetles as companions on my pathway in landscape ecology.

⁓

I've mentioned before how people and events can come together to influence the direction of one's pathways in unexpected ways. So it was with ants, but by way of birds.

In 1989, Enrique Bucher came to Fort Collins with his family to spend a sabbatical leave with our group. Enrique was an ecologist on the faculty of the Universidad Nacional de Córdoba in Argentina who was interested in birds and the Argentine Gran Chaco.

The Gran Chaco extends over northern Argentina and parts of Paraguay, Bolivia, and Brazil, spanning a gradient from humid tropical forests in the northeast to arid savannas and grasslands in the southwest. Enrique had conducted studies in the semiarid Chaco in Salta Province. As we talked about his work it became apparent that we shared an attraction to semiarid habitats and their birds. My interest in the Chaco grew. We saw an opportunity to collaborate, emphasizing the ecological effects of land uses—a landscape approach. This focus seemed to match the objectives of UNESCO's Man and the Biosphere (MAB) Programme, so we applied for funding. We would assess how land-use practices in the Chaco affected the distributions of species and biodiversity. Because our proposal didn't have a strong sociological component, we were not optimistic that it would be approved. But it was, for three years of support. Such are the tipping points that can send you off on a new pathway.

EPISODE: ANTS IN ARGENTINA

To kick things off, I flew to Córdoba. The evening that I arrived, Enrique and I were to meet his colleague, Carlos Saravia Toledo, over dinner to begin planning our project. It was my first inkling that customs in Argentina were different. We met early in the evening, strolled around the city to get acquainted, and talked about what we might do. By 9 p.m. I was hungry (it had been a long day); by 10 p.m. I was hungry and impatient; by 11 p.m., when we finally ordered dinner, I was thinking only of food. Late dinners were the custom in Córdoba. I was guilty of assuming that what was normal in the United States would be the norm elsewhere as well. It was a minor but valuable lesson in the importance of respecting and adapting to local customs when traveling to other countries. And keeping some snacks available as a backup.

Carlos was a plant ecologist who was also a landowner in the Chaco. He had established a reserve on his property, Los Colorados, where he managed cattle grazing in fenced paddocks to foster restoration of vegetation in areas that had previously been heavily overgrazed. His land use contrasted with that in the adjacent area, where cattle and goats were free-ranging and grazing was unmanaged. Typically, settlers in this area lived in *puestos* (posts) at water bores. Livestock concentrated their grazing around the water source—most goats and cattle stayed within 2 km, while a few ventured out as far as 4 km. The concentrated grazing created a gradient of habitat degradation, from highly degraded areas with little vegetation and compacted soils close to the *puesto*, to sparse grass and thorny shrubs in moderately degraded areas a little farther away. An area at Los Colorados that had been fenced for three years was moderately restored, resulting in heavy thorn-scrub growth and a light cover of forbs and grasses, while a highly restored area that had been fenced for eighteen years and ungrazed for six years had lighter thorn-scrub cover and a dense cover of grasses and forbs. These four land-use types represented points along a gradient of decreasing habitat degradation that was typical of the Chaco. It would be an ideal framework for our investigations, one that could have applied as well as ecological implications.

Enrique and I returned to Fort Collins and began to refine our thinking and assemble the equipment we would need. At about this time, Brandon Bestelmeyer joined my group to begin his MS degree studies. It was opportune timing. Brandon's interest was piqued by what Enrique and I had in mind. So he joined our Chaco team.

At this time, the "intermediate disturbance hypothesis" was much in vogue in ecology. The hypothesis posited that the species diversity of a group of plants or animals would be low in areas with little disturbance (where competitively dominant species would be able to exclude others) and in areas with high disturbance (where few species would be able to tolerate the disturbance). Diversity would be greatest in areas with intermediate disturbance, where there was enough disturbance to prevent competitive dominance by a few species but not so severe as to eliminate many species. Our four land-use types seemed to fit this model: highly degraded and highly restored at opposite ends of the disturbance gradient and the other two types intermediate. We could measure diversity at these points on the gradient. We had a hypothesis with clear predictions to test!

But diversity of what? We'd wait until we were on the ground in the Chaco to decide which group of organisms would provide the best test of the hypothesis, but it would probably be birds. So Enrique, Brandon, and I took off for Argentina to meet with Carlos at Los Colorados to begin sampling to determine the best measures of diversity on the land-use gradient.

Because there were few ecological studies in the Chaco to tell us what to expect, we began by exploring three possibilities: birds (of course), small mammals (which Brandon had studied as an undergraduate), and ground-dwelling insects (because there seemed to be a lot of them running around). Although we could use the visual sampling methods I'd used in the grasslands and shrubsteppe in the open areas around the *puesto*, the vegetation was too dense in the thorn-scrub areas to see the birds. Using machetes, we cleared a path through the vegetation to set up mist nets to catch birds. We caught a lot, many of them different species (and all new and exciting to a North American ornithologist!). So bird diversity in this habitat was high, but "high" was scarcely a term that would enable quantitative comparisons. We also saw many birds avoiding the mist nets. Reluctantly, we concluded that it would be difficult to sample bird diversity across the gradient in a consistent way.

How about small mammals? We set out a grid of small-mammal live traps (Sherman traps), as students in my group had done at Sevilleta, in Alaska, and elsewhere. Normally, we'd expect to catch a dozen or more individuals in one night's trapping. In the Chaco, we caught only one in three nights of trapping. If we had studied the literature, we would have known that small-mammal densities are generally much lower in semiarid South American habitats than in the North American places where we had worked. But we hadn't. There were simply too few small mammals in our area to generate meaningful measures of diversity.

That left us with ground-dwelling insects. We set out a grid of pitfall traps. The next day, the traps were full of squirming insects, most of them ants. It was decided—we would use ants to assess how diversity changed across the land-use gradient and test the intermediate disturbance hypothesis. Ants became the focus of Brandon's master's thesis project. He returned to Los Colorados twice more to assess seasonal changes in ant communities and document how functional groups of scavenging ants interacted over food supplied at bait stations. He spent two years sorting pitfall samples.

All of this required that we (actually Brandon) know the ant species we were dealing with. Ant identification isn't easy, and the ant fauna of the Chaco was poorly known. With the help of Roy Snelling (of the Los Angeles County Museum of Natural History), Brandon learned the nuances of ant identification. By the time our studies in the Chaco ended, he had identified 104 species (or morphospecies) of ants from thirty-four genera in our samples, certainly enough to see how diversity varied across the land-use gradient.

The patterns varied seasonally but were clearest in the summer (dry season) samples.[4] Ant species diversity was high in the highly degraded and highly restored areas, low in the intermediate moderately degraded and moderately restored areas. Exactly the opposite of what was predicted by theory. Once again it seemed to be our fate to come up with anomalous results.

This left us scrambling for an explanation (something that, by now, I was getting pretty good at). The answer to the anomaly involved looking not just at a measure of local diversity *per se* but paying attention to which species accounted for the diversity in which areas. In addition to the typical Chacoan species, the extremes of the land-use gradient were enriched by species characteristic of adjacent biogeographic regions. The highly degraded *puesto* area supported ant species characteristic of the hot, dry conditions of the Monte Desert to the southwest, while species widely distributed in subtropical and Amazonian regions to the northeast were common in the lush undergrowth of the highly restored area. Neither group of ant species was common in the intermediate areas, so overall diversity there was reduced. The ant communities reflected the broader position of the Chaco as a transition between adjacent biogeographic regions. To interpret the pattern of local ant species diversity we needed to think at a regional scale.

The Chaco was full of opportunities for ecological studies and its ants were intriguing. But this was Brandon's pathway, not mine. Enrique finished his sabbatical and returned to Córdoba. I resumed working with graduate students and Tom in the grasslands at CPER and writing and speaking about landscape ecology. Brandon continued to study ants, conducting his PhD research on a gradient of grassland and desert sites in Colorado and New Mexico. He documented how our perception of these habitats, based largely on their vegetation, differed from that of ants, which responded more strongly to soil characteristics. And scale was important—regional variations in ant faunas had important effects on local assemblages, as we had found in the Chaco.[5]

I had decided not to embark on the ant pathway. Another student's project, however, offered an opportunity for collaboration and provided a doorway to another possible pathway. I might well have followed this pathway, but in this case the door was slammed

shut and the pathway couldn't be followed. It's instructive to
see why.

EPISODE: TIME, ENERGY, AND SANDPIPER MIGRATION

In 1991, Adrian (he goes by Bubba) Farmer was a mid-career research
scientist with the U.S. Geological Survey (USGS) in Fort Collins when
he decided to fold his research into a PhD program and joined my
group of students.[6] Bubba came with his research project defined
and supported through his agency. He outlined an ambitious five-
year project to explore trade-offs between time and energy alloca-
tions in the spring migration of pectoral sandpipers from the Texas
Gulf Coast to the breeding grounds in arctic Alaska. Time is important because sandpipers cannot arrive in the North too early (food
would not yet be available) or too late (there would be insufficient
time to lay eggs and raise young before weather deteriorates in late
summer). Energy is important because birds must have sufficient fat
reserves to undertake the long migration and yet (for females) still
have enough energy to produce a full clutch of eggs. Sandpipers use
wetland stopover sites along their migratory route to stoke up on
energy. If females took too long to replenish energy stores or stayed
too long at a poor-quality site, they might not arrive at the breeding grounds with enough time to breed or enough energy to produce eggs. Reproductive success would suffer—there would be fit-
ness consequences.

Bubba wanted to investigate how differences in the distribution
and quality (food and energy supply) of stopover sites along the
sandpipers' migratory route might affect their migratory strategy,
something that could affect how wetlands were managed on a
regional scale. He used dynamic programming to model the time-
energy trade-offs,[7] but he also conducted field studies at several
locations, using ground surveys and aircraft to monitor the length
of time that radio-tagged birds stayed at a stopover site before

taking off to continue their migration. It was an intensive project, possible only because of the support from USGS and the Global Change Program of the U.S. Department of the Interior.

Bubba modeled the consequences of two alternative migration strategies: either stay at a stopover for a short time before moving on to another, in a series of short jumps (the "mover" strategy), or stay a long time at one location, accumulating more energy before undertaking a long, unbroken flight north (the "dawdler" strategy).[8] It turned out that which strategy was best depended on both stopover distribution and quality. In a landscape with closely spaced, high-quality wetlands, birds could follow various combinations of the two strategies and still arrive at the breeding grounds early enough and with sufficient energy reserves to achieve maximum reproductive success. However, as stopovers became fewer and more widely separated and quality declined, the optimal strategy converged on the dawdler alternative. Sensitivity tests showed that stopover quality was generally more important than stopover distribution. These results had some obvious land-use implications: it would be better to have fewer widely spaced but very productive wetlands than more stopover sites of lower quality.

The distribution and availability of wetland stopover sites, as well as their productivity, are affected by land use and climate. A small change in annual precipitation, for example, may produce a large change in wetland productivity. Bubba had originally designed his project to assess the impacts of global climate change on the management of stopover habitats. The coarse resolution of global-change models at that time prevented him from addressing this aspect of his project in much detail. By the time he had completed his PhD in 1997, however, the state of climate modeling had improved. Bubba's position with USGS was based in Fort Collins, so we discussed the possibility of continuing his work collaboratively. Future climate change was likely to make precipitation more variable in the Great Plains corridor through which sandpipers migrated. This could alter the productivity of the remaining wetlands as well as increase the loss of wetlands as they were drained and converted to agriculture. We could use updated global-climate models in combination with Bubba's models of shorebird energetics to project the consequences of future changes in stopover distribution and productivity. We submitted a proposal to NSF and waited. I was excited about the prospect of conducting research on birds again and embarking on a pathway addressing climate change.

Alas, it was not to be. The proposal was not approved. The reviewers liked the overall concept and were at least mildly enthusiastic about the field component of the research. Even though we proposed using the latest state-of-the-art global-change models, the reviewers felt that the models would not provide an adequate foundation for assessing changes in precipitation, and thus variability in wetland distribution and condition. The problem, in short, was that the global-change models that we intended to use were still too coarse to project wetland quality in sufficient detail. Our proposal was ahead of its time. Bubba and I could not wait for the state of global-change modeling to improve enough to provide what the reviewers thought was necessary. We went on to other things, Bubba working on other research projects with USGS while I continued on the landscape ecology pathway.

Neither ants nor sandpipers offered a pathway that I wanted to (or could) follow. Ants were better left to Brandon and Tom, who had a much better feeling for the organisms than I did, and funding to support further sandpiper research with Bubba was not likely to appear. The interests of other students, however, would eventually lead me onto another pathway, although it took a long time to develop.

20

PULLED ONTO A
SEABIRD PATHWAY

I'm a terrestrial ecologist. I've headed out into the open spaces of arid and semiarid places—grasslands and shrubsteppe—all my life. I really don't like water. I didn't learn to swim until I was five or six, when my uncle offered me fifty cents (a fortune at the time) if I could swim across the pool at the college in Bethany. I did, pocketed the fifty cents, and swam as little as I could after that. I grew up in central Oklahoma, far from any ocean. I didn't even see the ocean until I was in college.

But then I took a faculty position at Oregon State University, only fifty miles from the Pacific Ocean. Even though I headed inland to the semiarid shrubsteppe to conduct my research, it was only a matter of time until someone would show up wanting to do graduate work with seabirds. That someone was Mike Scott.

I was just getting started in Corvallis when Mike and his wife, Sharon, showed up at my office. Mike had undergraduate and graduate degrees in marine biology from San Diego State University and was attracted to Oregon State for his PhD because of its strength in oceanography. He was also interested in birds and wanted to combine the two interests. It sounded intriguing, but I was just beginning to explore research options in the shrubsteppe; dealing with marine systems would be a stretch. I was also reluctant to take on an additional graduate student. But Mike was motivated and insistent, and Sharon was extraordinarily perceptive (she was surreptitiously interviewing me while I interviewed him). I agreed to accept him (and she me), thinking that I could probably steer him toward working in the shrubsteppe. Instead, he opened my eyes to the allure of the ocean and marine birds, and in the process steered me toward a different pathway.

Mike saw what I was doing in studying resource partitioning among birds in grasslands and shrubsteppe and decided to apply the same conceptual framework to seabirds.[1] He spent three years documenting the food habits, foraging behavior, and reproductive activities of four seabird species breeding on Yaquina Head on the central Oregon coast. He conducted ship-based transect surveys to determine nearshore and offshore distributions. Eventually, he enticed me to join him on some of these surveys. Seeing murres, cormorants, and vast flocks of shearwaters at sea was thrilling. But it wasn't enough to make me forsake terra firma.

Mike's enthusiasm for seabirds, however, drew him into other projects, and me along with him. Gulls were abundant along the coast, as they are in coastal areas in most of the world. Earlier, Mike had conducted a survey along the Oregon and Washington coasts, finding that some gulls had characteristics of both glaucous-winged and western gulls—apparent hybrids. The epicenter of hybridization was at Destruction Island, along the Olympic Peninsula of Washington.[2] Mike suggested that we might dig into this more deeply. We recruited a new graduate student in my program, Wayne Hoffman, to join us, and with some funding from the American Philosophical Society we headed off to Destruction Island to document gull hybridization in detail.[3]

EPISODE: THE GULLS OF DESTRUCTION ISLAND

Destruction Island is a 12-ha island located 6 km off the Washington coast, southwest of the Quileute tribal community of La Push.

Following several shipwrecks, a lighthouse was built on the island in the late nineteenth century. It was maintained by lighthouse keepers until the 1960s, when the light was automated. The lighthouse was still in operation (with the foghorn blasting every thirty seconds, fog or not) when we were there in 1974, but no one lived on the island any longer. It was a mecca for gulls and other seabirds.[4]

We chartered a fishing vessel at La Push to take us to the island, with instructions to return for us a week later. We had adequate provisions and permission from the U.S. Coast Guard to stay at the lighthouse building, so we were set to record multiple measurements of gulls to document the extent of hybridization between the two "pure" species. We planned to capture gulls using a cannon net—a large mesh net attached to weights that can be remotely fired out of tubes (the "cannons") to entrap birds lured into the net area by food. We set up the cannons and net on a beach at low tide, retreated to hide behind the rocks, and waited. The gulls flocked into the food. We fired the net. It arched over the birds and fell, trapping some twenty-five to thirty angry gulls. We spent several hours recording eye color, eye-ring color, feather colors, leg colors, and various body size measurements—variables that could be used to quantify the degree of hybridness of individuals. We then waited for the next low tide to do it again, which we did.

But this time it didn't work. We waited until a fair number of gulls had flocked into the area that would be covered by the net. We fired the cannons. The net arced out over the assembled gulls, all of which immediately took flight and were gone by the time the net hit the ground. Zero captures. Clearly, the gulls knew what was coming. These were not the same gulls we had captured before, so they had apparently learned about the cannons and net by watching what had happened to their compatriots. Not to be deterred, we tried again. Same result. Now we were deterred.

Since we wouldn't be able to capture gulls to measure eye and plumage characteristics of individuals directly, we resorted to observing paired birds at nest sites through spotting scopes. This enabled us to determine the degree of hybridness or purity of mated individuals and to document their reproductive success.

Over half of the birds that we assessed were hybrids. Individuals tended to pair with birds having similar traits—hybrids with hybrids and "pure" birds with another of the same species (i.e., assortative mating). Unexpectedly, pure pairs hatched fewer eggs than pairs with at least one hybrid individual. Over time, this should lead to a swamping of the pure forms by hybrids. We suggested that assortative mating, combined with a low but steady level of immigration of pure types to the island, would counterbalance the greater reproductive success of hybrids and maintain a stable mixture of hybrids and pure types in the population (a conclusion we confirmed with a bit of modeling wizardry).[5]

When the time came to leave the island we scanned the horizon, looking for the fishing vessel from La Push. Nothing. Late that day we received a radio message: even though the weather was clear and sunny, rough seas had prevented the boat leaving the harbor. It would come tomorrow.

Tomorrow came, and then another, and still no boat. We began to feel like castaways. Our provisions were running low. We tried to fashion fishing hooks from safety pins but caught only one small rockfish. There were plenty of large rabbits on the island, descendants of a pet the lighthouse keeper's daughter had kept decades before. But we had no way to catch them. Finally, on the fourth day the boat appeared, and we were able to leave the island. We were no longer marooned with the gulls.

Wayne continued to survey gull hybridization along the Washington coast and returned to Destruction Island the next year, but I was glad to get back to working on the firmer footing and wide-open expanses of the shrubsteppe. I thought I was finished with seabirds and no longer on that pathway. But the students wouldn't let me off so easily. Although Mike Scott had gone on to a position with the U.S. Fish & Wildlife Service in Hawaii, Wayne continued to analyze gull data for his master's thesis, now joined by another student interested in seabirds, Dennis Heinemann. At that time, the federal government was launching a multiagency program to assess the potential effects of oil development in several lease parcels on Alaska's continental shelf (the Outer Continental Shelf Environmental Assessment Program, OCSEAP). Wayne and Dennis convinced me to apply for funding to support them, but it would also pull me back onto the seabird pathway.

We initially received support for one year to produce baseline information on how seabirds might use areas projected for leasing. We focused on the dynamics of mixed-species seabird foraging flocks. These flocks form over prey aggregations such as schooling fish and may include more than a dozen seabird species— shearwaters, puffins, auklets, guillemots, murres, loons, and cormorants as well as several gull species. The concentration of birds in feeding flocks would make them especially vulnerable to the effects of an oil spill.

Wayne and Dennis spent most of the summer observing flocks off Kodiak Island and in the Gulf of Alaska. I joined Wayne for one incredibly rainy week on Kodiak, sitting in the rain on headlands recording observations of nearshore feeding flocks. Our results were apparently sufficient to convince OCSEAP to continue supporting the research for another four years.[6] I recruited a postdoc, Glenn Ford, and our emphasis turned to modeling, building on the modeling of seabird energetics that Mike Scott and I had done several years before (see chapter 9). I was back on a combined seabird and modeling pathway.

Wayne decided to move to Florida to continue his graduate work, while Glenn, Dennis, and I developed models and gathered data to populate them. Fortunately, there was not an actual oil spill during our studies. Instead, we constructed several hypothetical spill scenarios that differed in magnitude, location, spatial extent, spread, and time in relation to seabird breeding activities. We then modeled the impacts of each spill scenario on populations of murres and kittiwakes breeding on the Pribilof Islands in the Bering Sea. The models were quite detailed (Glenn kept thinking of additional variables to include), considering such things as trip time and for-aging load for adults with and without chicks of different ages and energy demands. The modeling results were similarly detailed but revealing. The effects of a one-time oil spill on murre popula-tions, for example, were greater if adults rather than first-year birds suffered deaths. Small reductions in fecundity or survival due to low-level but chronic oil pollution, such as that from recre-ational fishing boats, could have greater long-term effects than larger, more immediate mortality from a massive onetime spill. Model projections were more sensitive to oil impacts on foraging distribution and food availability than to the probability of adult death from a spill.[7]

We published papers and presented the results of our modeling at scientific meetings. Our contract with OCSEAP completed, we moved on to other things. Glenn left to form his own environmen-tal consulting firm based on his modeling expertise. Dennis moved with me to UNM and completed a PhD study on the interactions between hummingbirds and wasps foraging on flowers in high-elevation meadows. And I moved to New Mexico and started down the landscape ecology pathway. The pathway to seabirds and oceans had been on-again, off-again, and then on-again. People and oppor-tunities had kept drawing me back. But now, once again, I thought I was finished with seabirds and the ocean.

But unforeseen events have a way of pushing you onto pathways that you had no intention of following. I didn't anticipate the *Exxon Valdez* oil spill.

21

DEALING WITH ADVOCACY
The *Exxon Valdez* Oil Spill

On March 24, 1989, the oil tanker *Exxon Valdez* ran aground on Bligh Reef off the coast of Bligh Island, Alaska. The ruptured tanks spilled eleven million gallons (forty million liters) of crude oil into the pristine waters of Prince William Sound. As the oil spread it was deposited in thick, gooey layers on the shorelines, and then along the Kenai Peninsula. The spill eventually affected some 1,300 miles (2,100 km) of shoreline.

It wasn't long before oiled seabirds began to wash up on shorelines. As carcasses accumulated, the magnitude of the mortality became clear. Within weeks, more than 30,000 oiled carcasses had been retrieved—only 12 percent of the 250,000 birds eventually estimated to have been killed by the spill.[1] Media coverage was intense because of the size of the spill, because the images of oiled shorelines and dead and dying birds were so evocative, and because a large oil corporation was the culprit. Exxon and state and federal agencies quickly mounted efforts to clean the oil from shorelines, rescue as many oiled birds and sea otters as possible, plan how to assess the extent of damages to biological resources, and determine whether and when recovery might be possible (and, of course, determine who was to blame).[2]

I had seen the news reports and was as dismayed and outraged as anyone. But it didn't occur to me that the spill would affect me personally, that it would draw me back onto the seabird pathway. But a few weeks after the spill I got a call from someone at Exxon. They were going to launch investigations of the effects of the spill and were looking for outside scientists to review research proposals. They had dug into the scientific literature dealing with seabirds and oil and had come across our accounts of modeling hypothetical oil spills on the Pribilofs. Would I be willing to spend a few days reviewing proposals, visiting the spill area in Alaska, and advising them on what sorts of studies they might undertake?

I was reluctant to become involved. Big corporations like Exxon had a reputation (deserved or not) for tilting science to serve their interests. But I thought I might be able to influence investigations, to hold them to account for what was clearly an environmental disaster. And it would provide an opportunity to see the spill and cleanup responses firsthand, something I could bring back into the classroom. So I read the proposals and went to Alaska to be ferried around the spill area by helicopter. It was even worse than I had imagined: shorelines turned black and glistening with thick layers of oil; morgues full of oiled carcasses of murres, cormorants, gulls, shorebirds, waterfowl, and other victims; massive flotillas of ships and thousands of people cleaning the oil from shorelines. I resolved to do what I could to ensure that follow-up research was scientifically sound, independent, and tough-minded.

The reviewers met with Exxon scientists and staff in Seattle to discuss what might be done to document the biological consequences of the spill. I made my recommendations, framed as strongly as I could, and returned to Fort Collins to share my impressions of the spill and Exxon's response with students and continue my work in landscape ecology. But a week later I got another call from Exxon, this time from a senior executive. Following my recommendations, they had decided to award a contract for seabird research to a Fairbanks-based environmental consulting firm, ABR Inc. Exxon wanted me to work with ABR scientists to develop an Exxon-sponsored research program on the effects of the oil spill on seabirds. This would entail more than incidental consulting now and then; it would involve my direct and ongoing participation in the research, in the field, in Alaska, and (as I was later to discover) serving as an expert witness in litigation over spill damages.

What should I do? I was suspicious of Exxon and wary of being associated with them. After all, Exxon had spilled the oil. They might use science to cover up their misdeed. I challenged them on what they expected of science and scientists. "Look," they said. "We know it's bad. We need to know just how bad it is based on the strongest, most defensible science. Tell us the facts so we'll know what we're dealing with."

This was not what I expected to hear. We talked over the details. The work would be done to provide Exxon with the information they needed—it would be contract research, in which the sponsor asked the questions. But it was also an opportunity for us to ask questions about the ecological responses of birds to such a massive habitat disruption and to assess our earlier modeling of oil spill effects on seabirds on the Pribilofs. Exxon would provide the support and funding, but it would be up to my team to design and implement the research. I was impressed with the ABR scientists who would lead the fieldwork, Steve Murphy and Bob Day, and thought we would make a good team. And it would get me back to working with birds again.

So I agreed. I expected that my involvement with the *Exxon Valdez* spill might last a year or two. It turned out that I would be following this pathway off and on for the next quarter-century.

Large environmental accidents such as the *Exxon Valdez* spill do not escape the notice of lawyers. Within days of the spill, multiple lawsuits had been filed against Exxon by multiple parties. There promised to be lengthy and contentious legal sparring.

The lawsuits dealing with damages to biological resources were brought by an assortment of government agencies (collectively, the "Trustees"). The positions of the Trustees as plaintiffs and of Exxon as defendant led each group to sponsor separate scientific studies with different teams of scientists. It was in the Trustees' interest to document damages, for which (as Trustees of the environment) they sought compensation. The presumption was that a species must have been negatively affected by the spill through direct mortality, reproductive failure, avoidance of damaged habitat, or all the above. The burden of proof was therefore on showing that a species had *not* suffered negative impacts. Because Exxon's focus was on defending themselves, their interest was not only in documenting which species had or had not shown spill impacts, but also in determining whether, and how quickly, the impacted species recovered. The burden of proof was on showing that a species *had* suffered negative impacts and that these impacts continued. Litigation polarized the scientists and drove them to ask different questions, use different approaches, and interpret data in different ways.

The legal setting of our work was something new to me. Exxon made it clear that they expected us to publish our results in top scientific journals, so our work would need to meet the standards of peer reviewers who might be predisposed to be skeptical of our findings. Our findings would also need to withstand the intense scrutiny of the public, media, other scientists, and sharp-eyed lawyers, all intent on finding flaws in our studies or casting doubt on our conclusions. Litigation set the context in which Steve, Bob, and I designed the seabird studies. With the help of a statistician, Keith Parker, we adopted a statistical approach, defining an impact as a statistically significant relationship between a biological measure (e.g., the abundance of a species) and a quantitative measure of exposure to oil. Recovery could then be defined as the disappearance over time of a previously documented statistical relationship with oiling.[3]

In my previous work I had documented the importance of temporal variation in grasslands and shrubsteppe, and I thought that conditions in Prince William Sound might be even more variable. Detecting recovery would therefore require multiple surveys over time to separate spill effects from background environmental variation. When we outlined our research plan to a group of senior Exxon executives, however, they questioned our approach. "Why do you need to go back and survey again if you have already done so?" they asked. Their backgrounds were in petroleum engineering, so they were used to dealing with geological systems that don't change (at least on a human timescale). For such systems, one-shot sampling may be all that is needed. I fell back on a class lecture to show them how ecological systems were different. They were sharp people and immediately got it. Some of them turned out to be the most perceptive, critical, and helpful reviewers of the science that we produced.

To detect oiling effects statistically, we had to be able to relate seabirds to a quantitative measure of oiling at an appropriate scale. Oiling of shorelines and nearshore waters had obvious effects on the habitats available to birds. Yet neither visual appearances nor detailed measurements of hydrocarbon concentrations in beach sediments could tell us how birds might perceive the suitability of their habitat after such a massive perturbation. We needed to couple measures of oiling exposure with an appropriate measure of the birds' responses to oiling of their habitats. In ten bays in Prince William Sound we quantified the exposure of shorelines to oiling, expressed as an oiling index. Because the bays differed in habitat features as well as oiling, we also measured a set of habitat features relevant to seabirds in each bay. We could then assess counts of birds in each bay in relation to the oiling index, with and without the habitat measures. This allowed us to determine whether differences in abundance were due to oiling, above and beyond the underlying differences in habitats among bays. The inclusion of habitat measures was a manifestation of my long-term interest in habitat measures.

We conducted surveys using an inflatable fourteen-foot zodiac launched from a "mother ship" (usually an eighty-five-foot fishing vessel) anchored in a bay. The zodiac was driven slowly along the shoreline while two observers identified and counted all birds on the shore and on the water along the shoreline. The bays were large with convoluted shorelines, so it took a day or more to survey each one. We repeated surveys in each bay two to five times during multiple sampling periods in 1989 through 1991 and intermittently for several years thereafter.

It was a massive effort, not without its moments. Once, the fog rolled in and we lost sight of the shoreline. We tried to navigate with no landmarks and no idea where we were (this was before GPS was routinely available). We were lucky. We had been headed toward the headwater of the bay instead of its mouth, where we would have gone on into the open expanse of Prince William Sound. Another time we were in the zodiac going from one bay to another and found ourselves in the shipping lane for tankers leaving Valdez. A gigantic tanker was bearing down on us. Figuring that we were no match for the massive vessel, we quickly changed course and sped to safety. Then there was the time our outboard motor failed, and reverse gear was all we had. After trying to steer the zodiac backwards and only going around in circles, we eventually paddled to shore and waited to be rescued by another skiff from the mother ship. Fortunately, it was a matter of hours, not days.

Between surveys we often passed time fishing from the mother ship. I seemed to have a knack for catching halibut, perhaps because the skill required involved little more than baiting a hook, tossing it overboard, letting it sink to the bottom, and then slowly jiggling the bait up and down. Somehow, I attracted the attention of a giant, 350-plus pound halibut. I tried to reel it in while the fish did its best to stay on the bottom. It was a contest to see who tired first. After nearly an hour, I won. We needed a winch to bring it on board the mother ship. When I returned home to Colorado and our daughters asked what I had brought them from Alaska, I could tell them there was fine fish for dinner (and a great fish story to go along with it). I think they had hoped for something better.

When we put all the survey counts together and ran the statistics, the results were clear.[4] Initial spill impacts and subsequent recovery varied hugely among species. Some species, such as red-necked phalaropes and pigeon guillemots, suffered no apparent negative impacts to their occupancy of oiled habitats. These results were not too surprising; phalaropes occurred in the open water of the bays where little oil remained even soon after the spill, and guillemots occupied talus slopes and cliffs where little oil was deposited. Other species that stayed close to the shoreline, such as buffleheads or horned grebes, continued to exhibit negative associations with the oiling index values for several years. The proportion of species showing negative impacts steadily decreased over time, from more than half, two months after the spill to only a tenth, twenty-eight months later. Of the twenty-five species we were able to evaluate statistically over the duration of our surveys (1989–2001), twelve never exhibited a significant spill impact, ten showed an initial negative impact that later disappeared, and three showed a significant positive relationship with oiling that subsequently disappeared.

These findings contrasted with those of the Trustee-sponsored studies. We found even more species that showed initial negative impacts of oiling, but we also found more rapid recovery of most species. Only by 2013 did the Trustees declare that the last seabird still impacted by the spill, harlequin ducks, had recovered. When we considered oiling levels alone, we concluded that harlequin ducks were initially negatively impacted but recovered by 1996. When habitat variables were included in the analysis, however, all indications of negative impacts disappeared. The initial results were apparently due to habitat differences among the bays, not oiling.

The differences in results fueled contentious arguments about who was right. The different conclusions, however, stemmed largely from differences in how the questions were initially framed and what then followed. How the studies were designed, which species were considered, how impact and recovery were defined, and the methods of data analysis—all traced back to the initial emphasis on documenting damages versus detecting recovery. Unfortunately, these differences fueled a public perception of "dueling scientists," scientists who toss claims and counterclaims back and forth. This affected the credibility of all scientists studying the spill, regardless of which side they were on.

As the litigation over damages from the oil spill continued, our work to document any lingering impacts on seabirds and evaluate their recovery also continued. We were unable to detect any statistically significant effects of oiling on habitat occupancy by any of the species we were able to analyze in our 1996 survey.[5] But there were still isolated pockets of oil sequestered under the surface of shorelines in a few areas sheltered from natural weathering. Some researchers had speculated that these deposits might be released by disturbance (e.g., by sea otters digging for prey) and have delayed effects on species such as harlequin ducks. So we conducted another set of bay surveys in 2001. As before, we found no evidence of lingering oil impacts.[6] This was our last survey in Prince William Sound. Years later you could still find small deposits of oil if you knew where to look, but the oil was no longer toxic and would remain sequestered without vigorous digging.

I continued to navigate this oil spill/seabird pathway intermittently for several more years, analyzing data, publishing papers, giving presentations, and talking with lawyers now and then. Mostly, however, I had returned to the landscape ecology pathway, and then headed in a new direction (as I will describe shortly). In 2010, as the litigation was drawing to a close,[7] the Exxon leadership decided to sponsor a series of scientific review papers to summarize the findings about the immediate and long-term effects of the spill on fish, invertebrates, birds, and sea otters. Steve, Bob, and I were asked to write the paper on seabirds. I felt that the impact of these papers might be greater if they were synthesized and published as a book rather than separate papers scattered among different journals with specialized readerships. My suggestion made its way to the Exxon leadership, who gave it the go-ahead (and necessary support), *provided* I would edit the volume. I agreed, only to realize later that this would pull me back onto the oil spill/seabird pathway for another year or two. It was an opportunity to work with top scientists in several disciplines to integrate their work and impressions into a cohesive treatment that was much more than a recounting of who found out what and when. Cambridge University Press published the resulting volume, *Oil in the Environment: Legacies and Lessons of the Exxon Valdez Oil Spill*, in 2013.[8] That's when I was finally able to leave this pathway, decades after Mike Scott had first pulled me onto a boat to survey seabirds off Yaquina Head.

Scientists often take different approaches to the same problem and disagree about their conclusions; that's part of how science operates. However, pressure to take the science in a particular direction—to come up with the "right" answer to a question—can easily lead one to advocacy. The line separating science from advocacy is thin. Sooner or later, you may be drawn into a situation where advocacy threatens science, and it's not always easy to separate the two. As an ecologist, you may be asked to comment on the ecological costs and benefits of environmental actions, such as logging, draining a wetland, or grazing a grassland (or an oil spill). Because ecological systems are complex, variable, and subject to scale effects, there are rarely simple "yes-or-no" answers to such issues, but rather "it all depends." When the actions involve people, societal values can create additional conflicts. The pressure to use science to support one position or another can become intense, increasing the peril of slipping into advocacy. The best you can do is to critically examine your actions and assumptions, be true to your data, and remain alert to the creeping reach of advocacy. Still, even the way the questions guiding your research are posed can unintentionally tilt the research toward advocacy of a particular conclusion. Whether the initial expectation following the *Exxon Valdez* oil spill, for example, was of no effect or of a negative effect determined the design of studies, the statistical tests used, and the conclusions reached.

My experience working with Exxon on the oil spill aftermath reinforced my long-standing beliefs that considerations of habitat and spatial and temporal variation were vital to interpreting ecological patterns, whether they be natural or the results of environmental disturbance. It showed me that it is possible to combine rigorous scientific research with environmental consulting *if* one holds fast to the principle of objectivity and avoids advocacy. Consulting can provide an opportunity to do real science, but one must beware of attached strings.

Exxon emphasized the importance of openness and transparency in our work. They placed no restrictions on our publishing our results and insisted that we publish *everything*. This was unusual and unexpected. Some years later I was contacted about conducting similar studies of the effects of the *Deepwater Horizon* blowout and oil spill on seabirds. I was sent a consulting contract to sign. The contract stipulated that none of the work could ever be published. I threw the contract away.

In the introduction, I noted several factors that can determine one's route along a pathway. My involvement with seabirds began by investigating gull hybridization and advising Mike on his research into competition and bird community organization. It was basic research, driven by key *people*, my *interests* and *motivation*, and *opportunity*. Our findings were perhaps interesting to our scientific peers, but they had no obvious practical applications. As my journey along the seabird pathway developed, however, the work began to address questions that people other than scientists cared about. It became applied research, designed to use scientific findings to inform decisions that affected people as well as nature—to inform *societal concerns*. This is the role of ecologists in areas of natural resource or environmental management, such as wildlife management, fisheries, habitat restoration, or forestry. Or conservation, as I was soon to discover.

22

SHIFTING PATHWAYS TO CONSERVATION

onservation lies at the intersection of basic and applied science. The goals of conservation are to protect and preserve species, biodiversity, and ecosystems and the places they need to persist. Success rests on an understanding of such things as the dynamics of diminishing populations, patterns of habitat occupancy and resource use, interactions among species, and the environmental tableau of ecological niches. These are the domain of basic ecological research. But conservation is also about using a foundation of ecological concepts and principles to inform actions—applied science. To mount effective conservation efforts requires both.

Conservation is also an advocacy science. Most ecologists, and all conservationists, are advocates for nature—for at-risk species and biodiversity and against the forces of humanity that would destroy nature. To them, this is a fundamental ethical and moral position. Philosophically, they may align with those who argue for the value of nature in its own right—its intrinsic value—or those who promote nature for its usefulness to humans—its instrumental value.[1] In either case, however, nature is what they study and value.

As conservation biology came of age in the 1970s, and awareness of the impending biodiversity crisis deepened, ecologists began to think more about how their science could contribute to the conservation of imperiled species or the management of natural resources. I was no exception. I had written intermittently about what ecological science could contribute conceptually to help address broadly defined conservation problems. The fragmentation of habitats into small, isolated remnants, for example, was an issue facing conservationists in many parts of the world. When we were at Oslo in 1989, I was invited to address the Congress of the International Union of Game Biologists in Trondheim and write a paper on the linkage between habitat fragmentation and wildlife populations.[2] A few years later, I wrote another paper viewing fragmentation in the context of landscape mosaics, and then was invited to contribute chapters on habitat patchiness and conservation to several conservation or wildlife-management books.

My aim in these writings was to show how ideas about spatial patterning and dynamics from ecology and landscape ecology could enrich conservation thinking and practice. I was approaching conservation and resource management from the perspective of basic science, leaving it to others to see how (or even whether) the ideas could be applied. Although I was interested in conservation problems and supported conservation organizations, I did not think of myself as a conservation biologist. Except when work on the *Exxon Valdez* oil spill intervened, I was still firmly embedded in the landscape ecology pathway. The classes I taught at Colorado State were about landscape ecology; conservation teaching was in the Department of Fisheries and Wildlife.[3] But the colleagues who invited me to conferences or asked me to write book chapters were slowly but surely pulling me toward conservation.

The doorway to a new pathway in conservation opened in 2000–2001, when Bea and I took sabbatical leave from our positions at Colorado State. As we cast about for ideas about where to go and what to do, I had an opportunity to apply for a fellowship at the National Center for Ecological Analysis and Synthesis (NCEAS) in Santa

Barbara, California. NCEAS is an ecological think tank, drawing together scientists from varied disciplines to integrate information and ideas to address outstanding ecological problems. It seemed a perfect fit with my goal of strengthening the linkage of spatial dynamics and landscape ecology to both basic and applied ecology. I was not yet thinking about how to integrate landscape thinking with conservation.

My application to NCEAS was approved, providing the financial support to make moving to Santa Barbara for a year feasible and affording the time and intellectual atmosphere to think. Bea and I envisioned polishing off large backlogs of papers from our research programs and I would continue pushing the conceptual relevance of landscape ecology. We drove west to California, rented a house only a ten-minute bike ride from NCEAS, enrolled Taryn in the nearby school (Kyra was away at school in Massachusetts), and settled in to enjoy a relaxing and productive year.

But then an unforeseen opportunity came along, one that would launch me on the conservation pathway. Before we even got started on our sabbatical plans I got a call from Francis James. Fran was a fellow ornithologist whose interests were like mine. We had shared papers, talked about our projects when we'd see each other at meetings, and kept in touch for many years. Fran was on the Board of Governors of the Nature Conservancy (TNC). TNC was one of the world's largest conservation organizations, with a particular focus on procuring and protecting habitats ("Saving the Last Great Places" was their mantra at the time). TNC recognized the importance of basing their work on good science—over a hundred staff had advanced degrees in some relevant discipline. Because TNC planned and implemented conservation actions, virtually all of these scientists were at the applied end of the basic-applied spectrum. Fran was concerned that they might not be able to bring the latest findings of basic science (i.e., ecology) to bear on the conservation issues they faced. She convinced the TNC board to support an independent review of science in TNC, and she was now asking me to lead this review.

My knowledge of TNC at that time did not extend beyond the phrase, "send us money" (which I did) "and we will buy land."[4] I knew nothing of their science programs. But a sabbatical leave releases you from the many duties of academic life that break productive time into small chunks. I was feeling this release from obligations and anticipated having some free time. I was also interested to see what science TNC had hiding behind the curtains of their public appeals for donations to buy land. And because Fran was a friend, I told her I'd do it.

With the help of TNC staff, I pulled together a review team of seven highly respected scientists.[5] For several months we conducted interviews, visited TNC offices and field sites, and tabulated the results of a survey of TNC science staff. We were impressed with the scientific knowledge and dedication of individual staff, but we saw that science was not well integrated into the work of the organization. Many science staff were involved in conservation planning, but few were engaged in actually conducting the studies to inform conservation actions. They found it hard to keep up with advances in their field. They often felt isolated from their scientific peers and even from a good deal of what TNC was doing. The problem seemed to be less a matter of science capacity than a consequence of how science was organized and used in such a broadly decentralized organization.

As we were developing our conclusions and recommendations, I was asked to present our preliminary thinking to a board meeting at TNC headquarters in Arlington, Virginia—in the Washington, DC, area. The TNC board included high-level corporate CEOs, major donors, and other "Very Important People"—the sorts of people that academic faculty like me rarely encounter. Somehow, I overcame my nervousness to summarize what we had done and where we were headed. Steve McCormick, who had just come on board as the new CEO of TNC, later cornered me in the hallway. He was enthused by my report and wanted to keep engaged as our team finalized its thinking. We agreed to keep in touch.

Steve was already on the job at TNC headquarters in Arlington, but his family was still in California, so he flew back every week or two. We arranged to meet in San Francisco so I could update him on our progress. On one occasion, we were discussing how science might be organized more effectively in TNC. We were (literally) sketching organizational charts on a paper napkin. At one point Steve pointed to a science leadership position and said, "Someone like you would be good at this." I don't remember how I replied, and our conversation moved on to other things. But a seed had been planted.

I returned to Santa Barbara and discussed the idea with Bea. She had been feeling burned out by her academic job and was thinking about trying something different. It hadn't occurred to me that I might leave academia, which had been part of my life since childhood. But the idea of bringing science into a large organization that was actually *doing* conservation was attractive. So I got in touch with Steve to pursue the ideas that we had outlined on the napkin. We talked some more, and a few weeks later I was offered a position as one of three lead scientists—a triumvirate that would be responsible for implementing the recommendations of our review and integrating science more fully into the work of TNC. Some of these recommendations—emphasizing landscapes, recognizing the importance of scale, confronting environmental variation and complexity—were themes that had run through my own research for decades. TNC would provide an opportunity to put this experience into conservation practice.

Why would I even consider leaving academia to set out on an entirely new pathway? I was comfortable where I was. I held a distinguished professorship at Colorado State, which freed me from some of the routine tasks that often weigh faculty down. I could attract outstanding graduate students. I enjoyed my work in landscape ecology and felt that I was having an impact on the development of the discipline. I also had considerable inertia, as many senior professors often do. But the challenge of putting my experience to use in advancing the cause of conservation was intriguing, and the opportunity to instill some of my thinking into the work of a major conservation organization was enticing.

Of course there were also personal consequences to consider. Bea and I had lived in Fort Collins for seventeen years, in the only home our children had ever known. We loved Colorado and our cabin in the Rockies. If I joined TNC we would need to move to the Washington, DC, area. We would have to adjust to living on the East Coast, with no wide-open spaces, no real mountains, and lots of traffic. And what would Bea do? How would this move affect her career?

By a remarkable coincidence, Bea had just been contacted by a former colleague asking whether she might be interested in a position as Program Leader for Wildlife with the U.S. Forest Service. It would be a good fit—Bea's research was closely linked with wildlife management, and she had previously collaborated with the Forest Service. Not only that; the position would be located in the Forest Service offices in Arlington, not far from TNC headquarters. Perhaps it really was time for a change.

We decided to accept the offers, to leave Colorado, the university, and our comfortable professorships to focus more directly on applying rather than generating ecological knowledge. We were both ready to follow new career pathways.

Consider for a moment how forces converged to shift me onto the conservation pathway. *People* provided the impetus, as is often the case with pathway shifts. Fran James triggered my digging into the inner workings of science in a conservation organization, and Steve McCormick created the *opportunity* to shift. The timing was propitious. I was on a sabbatical and my mind was open to new possibilities. Steve was exerting new leadership in TNC and looking for new ideas and fresh faces. I would have a hand in implementing the recommendations of our review instead of seeing our report gather dust as "shelf art." And there was the *location*. How likely is it that two tenured full professors, married to each other, would consider changing their pathways and lives to end up halfway across the country in exciting new jobs located only a few Metro stops from each other?

Faculty members on sabbatical leave are often tempted by other possibilities—greener pastures beckon. To counter such temptations, universities usually attach a rider to sabbatical agreements: you are obligated to return to the university for at least an academic year before jumping ship. We returned to Fort Collins when our time in Santa Barbara was up to resume our academic duties. Neither TNC nor the Forest Service wanted to wait a year for us to show up, however. We both taught classes and worked with students during the fall term but began our new jobs in January. Bea moved to Arlington while I stayed in Fort Collins so Taryn could finish the school year. We bought a house in a suburb of DC, united in May, enrolled Taryn in Georgetown Day School, and began adjusting to living on the East Coast.

23

BOLSTERING CONSERVATION SCIENCE IN THE NATURE CONSERVANCY

I joined the Nature Conservancy (TNC) as a lead scientist in 2002. Together with the two other lead scientists, Peter Kareiva and Sanjayan,[1] we were the new science leadership for TNC. Our primary goal was to bolster science and integrate it into the fabric of the organization. This involved breaking down the existing partitioning of science as a separate program and instead promoting science as a contributor to all aspects of TNC's work. We worked with TNC's large philanthropy program to infuse science into fundraising efforts and with communications staff to make science part of the public face of the organization. We wanted people to know that TNC didn't buy just any lands or "pretty places" for conservation; it did so based on a rigorous scientific assessment of the conservation gains. Science was the foundation of the ecoregional planning system that TNC had developed to target and prioritize where conservation investments would have the greatest impact.

Strengthening science in TNC also entailed enhancing the participation of their scientists in the external scientific community. We encouraged staff to publish their work in scientific journals and attend science conferences, and we held several writing workshops to help staff prepare and submit papers for publication. We organized internal TNC science conferences in which staff could spend several days sharing their work and ideas with other TNC scientists and managers. And we visited TNC scientists where they worked and brainstormed how to address both the scientific and organizational challenges that they faced.

Much of this was similar to what I'd been doing with students for decades. But bolstering science in TNC by empowering their scientists was only part of the job. TNC scientists were already convinced of the importance of basing conservation actions on good science; they only needed encouragement and recognition of their efforts. Much of what we did entailed selling science to the board and other leaders and managers in TNC. Several board members were CEOs of large technology businesses and were staunch advocates of bringing science into the mainstream of TNC activities. On the other hand, program and division leaders in TNC had to deal with specific TNC actions and challenges on the ground and were primarily interested in what science could do for them when it supported their goals. They were less enthusiastic if it looked like science would restrict what they wanted to do.

One of TNC's core activities, for example, involved transactions to buy land for conservation—protected areas of various sorts. Many of these transactions came through large donations or bequests, which often entailed lengthy negotiations. Such big deals were an important measure of success for the individuals who brokered the deal. Transactions costing more than $1 million required approval by a group of TNC leaders. I represented science in this group. Usually, the biodiversity benefits of a project were clear, and the discussions centered on financial issues. Occasionally, however, science came into play. On one occasion we were asked to approve a proposal to purchase a tract of low-lying coastal property that had obvious conservation value. Only a week before, however, I had seen a presentation by a TNC scientist describing scenarios of sea level rise in the same area. The sea level projections showed that there was a strong likelihood that the sea would inundate the tract in a few decades, affecting its conservation value. When I presented my concerns to the group they didn't discount the potential problem. But the analysis was based on model projections, which were uncertain. Moreover, the opportunity to buy the tract was immediate—now— and sea level rise was off sometime in the future. The proposal was approved. It was a clear example of the difficulty of balancing immediate gains against the possible long-term costs of sea level rise and climate change.

I was also concerned that some of the places TNC protected for conservation might no longer protect what they were intended to protect as the effects of climate change played out.[2] Although the places themselves would still be protected, the biodiversity and conservation value of those places would change as species came and went. I discussed the issue with a few senior managers. "We get it," they said. "But don't just tell us the problem. Tell us what to do." I realized then that TNC was primarily an action organization; scientific arguments were important insofar as they supported direct actions. Donors and supporters expected quick results. Hypothetical arguments about what might happen in the future carried little weight.[3]

Of course, TNC was not the only player on the conservation stage. In addition to the federal land- or resource-management agencies headquartered in Washington, DC, several large conservation NGOs (nongovernmental organizations) were based in the area. Most had missions complementary to that of TNC, and some had senior scientists in positions like ours. The potential for collaborations seemed obvious to me, since so much of my previous research had been built on close collaborations. So I met with my counterparts now and then to discuss possibilities (generally during working hours over coffee, sometimes after work over beers).[4] These encounters were usually open and friendly (after all, we were scientific colleagues), but sometimes there was a strained undercurrent. The NGOs often approached the same donors for major support, so we were competitors despite our shared goals. On several occasions, however, we found an avenue for successful collaboration.

One such collaboration developed from conversations I had with Tony Janetos, an ecologist who directed the Global Change Program at the H. John Heinz III Center for Science, Economics and the Environment in Washington, DC. We shared similar interests and got to talking about ecological thresholds. Thresholds could lead to abrupt changes, moving an ecosystem from one condition in which management or conservation practices worked to another in which they suddenly didn't. Thresholds were likely to become more frequent with climate change (his interest) and complicate conservation efforts (my interest). With joint support from TNC and the Heinz Center we organized a workshop, brought several dozen scientists to the Airlie Center in northern Virginia, and discussed thresholds for several days. I'm not sure that any great insights emerged, but everyone's thinking about thresholds and their implications was sharpened, and Tony and I took the messages from the workshop and tried to work them into the practices of our organizations. Again, however, I was met with the "I get it, tell me what to do" response from TNC managers. Many science staff now approached their work with a heightened awareness of the nasty consequences of unanticipated thresholds.

On other occasions we collaborated with government agencies. Although the agencies (like TNC) focused on the applications of science more than its conceptual foundations, sometimes we were able to combine both. At one point I was sharing thoughts with Greg Hayward of the U.S. Forest Service about the effects of temporal environmental variation on resource management (his interest) and protection of places for conservation (my interest).[5] Our focus was on the concept of Historical Range of Variation (HRV), the idea that ecological systems change over time within a limited range of variation determined by their past history. HRV provides a way of recognizing that management actions take place in a variable environment. Greg was concerned that the concept was being mis-applied in forest management. I saw HRV as a way of incorporating the environmental variation that had long interested me into con-servation planning and practice.

With support from TNC and the Forest Service, Greg and I, together with two other U.S. Forest Service scientists, Hugh Safford and Cathy Giffen, organized a workshop that brought together more than forty scientists representing a spectrum of basic and applied areas. It became clear that the three elements of HRV—history, range, and variation—could compromise both resource manage-ment and conservation. "History" meant that any system bore the imprint of what it had experienced in the past, so it would be risky to assume that similar places had the same history and could be treated in the same way. "Range" implied that there was a fixed envelope of conditions within which a system could be managed or conserved. And "variation" meant that assuming a system would remain unchanged from one time to another would be especially risky. There was a close relationship to the thresholds workshop: changes that fell outside of HRV because of environmental changes or management practices could push a system across thresholds, complicating management and conservation. Both workshops emphasized ecological concepts that had a direct bearing on appli-cations, and both had a common underlying theme: the times are changing, and the changes will affect what we do.[6]

Infusing science into the work of TNC involved much more than applying concepts and approaches from basic science to conserva-tion actions. Conservation is carried out in a societal context, which creates additional complexity, imposes significant constraints, and cannot be ignored. *People* were becoming an important feature of TNC's evolving conservation agenda and figured prominently in the organization's marketing efforts. *The Nature Conservancy* magazine, which had previously featured stunning photographs of places, plants, or animals on its covers, now began to highlight people. The message was clear: people are not only a threat to nature but are also a part of the nature that needs to be conserved. Conservation cannot proceed as if people don't matter.

As TNC's conservation programs expanded to include people as well as places, the organization changed in other ways. Although the state and country programs continued to focus on the conservation of places and biodiversity, activities at the headquarters in Arlington (the "worldwide office") reflected the status of TNC as a large corporation (albeit a nonprofit one). For TNC, conservation was a business as well as a mission; its annual cash flow exceeded a billion dollars. The feasibility of a transaction to protect an area often depended as much on financial arrangements as on biodiversity gains.

The economic side of conservation affected the activities of the three of us as lead scientists. Sanjayan, blessed with an abundance of charisma, became more active in philanthropy efforts, selling TNC's science-based conservation to major donors. Peter Kareiva put more of his time into developing collaborations with scientists and economists from Stanford and other universities to assess the "natural capital" of ecosystem services—the value of the benefits that natural ecosystems provide to people, such as clean air and water or insect pollination of agricultural crops. Ecosystem services provided a way to sell conservation to people who didn't see a moral responsibility for preserving nature for nature's sake but might value nature for what it contributed to their own well-being. Meanwhile, TNC underwent another reorganization. I was promoted to chief scientist, largely I think, because I was the only lead scientist who was present at TNC's worldwide office on a daily basis. I became more involved in conducting meetings, empowering staff, helping with outreach and communication, pulling together internal and external scientists to address issues (such as thresholds), and helping sell TNC through its science.

My pathway into conservation with TNC, however, was not what I had expected. I was helping to bring science from ecology and landscape ecology into the planning and practices of TNC, as I had hoped. But science in TNC was applied to answer the questions TNC managers were asking to support conservation actions. Where should we invest in land? What is the biodiversity of this place? How does protection of this place complement the conservation benefits of the places we've already protected? What can we do to protect this species? How will protection of this place or this species help advance our fundraising efforts so we can do more? My role was to administer the science to address such questions. It was exciting to be so closely involved in the dealings of a large, multinational, action-oriented conservation organization, and I felt that I was helping to put conservation on a firmer scientific footing. But I missed being more directly involved in the actual conduct of science.

Meanwhile, Bea, Taryn, and I were living on the East Coast, in the nation's capital, far from Colorado and the wide-open spaces of the West. There was certainly plenty to do in the Washington, DC, area. We went to plays with first-rate casts, held season tickets to the National Symphony, attended performances of the Washington National Opera at the Kennedy Center, and frequented a variety of events at the Birchmere (the area's fabled music hall) and a local coffee house, Jammin Java. We often visited the various museums of the Smithsonian. And we did our best to expose ourselves to the array of ethnic restaurants in the area.

We also spent some time exploring the surrounding area, especially its history. The region had been hotly contested during the American Civil War (1861–1865) and there were many battlefields to visit. Manassas (Bull Run), where Union and Confederate forces had fought two intense battles early in the war, was only a short distance from our house. Antietam, where some 23,000 soldiers were killed, wounded, or missing after only twelve hours of fighting, wasn't much farther.[7] We could easily picture troops traversing the area that would later become our backyard in movements during the war.

We didn't need to go far into the nearby Virginia countryside to see that the Civil War, while long over, was not forgotten. A shop along the roadside a few miles away sold all manner of Confederate memorabilia.

Despite the many attractions of the Washington, DC, area, our daily life was dominated by jobs and commuting. Taryn's school was a forty-five-minute commute from our house, followed by another half hour to our jobs (depending on traffic, which varied from just bad to terrible). We split the commutes, but they were a far cry from the bike trips to NCEAS in Santa Barbara or from our house in Fort Collins to the Colorado State University campus. The excitement and allure of our jobs and living in the DC area began to wear thin.

I was still committed to following the conservation pathway, but I was no longer sure that a science administrator position with TNC was the best way to do this. I missed the intellectual excitement of simply finding things out. Even though I knew my work at TNC was helping to advance science-based conservation, it wasn't satisfying. I guess my interests were still rooted in the basic sciences of ecology, landscape ecology, and ornithology and in developing concepts more than in generating data.

TNC was also undergoing more organizational change. Steve McCormick, who had championed our work as lead scientists, had left. Steve had come up through the organization and was imbued with its culture. The new leadership came from the world of business and investment rather than hands-on conservation. Although this may have been appropriate because TNC was now in the business of conservation, it changed the feeling of the organization, at least in the worldwide office.

The changes in TNC affected me directly. Several months previously I had shifted back into a lead scientist position, and now that position was eliminated as part of the reorganization. I could continue through the completion of the HRV workshop, but then my job at TNC would disappear, and I along with it. This prompted me to reflect on my journey on this pathway and it became apparent that I was not particularly well suited to the role I had been asked to play.

I probably paid too much attention to enhancing the basic science behind conservation and encouraging the scientists to engage with mainstream science, while failing to understand fully the roles of science and of TNC scientists in supporting the broader mission and actions of TNC. I wasn't good at forging the personal relationships with donors that were needed to promote TNC through its use of science to support conservation. I felt awkward talking with business leaders and didn't do a very good job of selling TNC's science.

In the end, I was probably still too much a professor. I could only go so far in shedding my academic robes. I had avoided administrative positions during my academic career, but now I found myself administering science as part of a business. It seemed to be time for a change.

24

DOING CONSERVATION SCIENCE

Although I hadn't planned to shift directions on the conservation pathway, the reorganization at TNC and the disappearance of my position left me little choice. I was open to other opportunities. One day I saw a notice on an ornithology listserv. PRBO Conservation Science in California was looking for a chief conservation science officer and wanted suggestions. I had known of PRBO (under its former identity as Point Reyes Bird Observatory) for years and had once visited their bird-banding station at Palomarin, near Point Reyes. The banding operation was now directed by Geoff Geupel, who had endured a year sorting insects in my lab as part of the sagebrush chemistry project I described earlier. So there was a personal connection.

Bea and I discussed the possibilities. Both of us were ready to escape the hectic pace of life and the long commutes in Washington, DC, and return to the West. Our daughters were both now off in school and we thought we could cope with the two of us being on opposite sides of the continent, at least for a while. Bea could continue her work with USGS in DC. After all, there were frequent (long) transcontinental flights between the East and West Coasts, so we could arrange to spend a weekend or two together every month. I applied for the position, flew to California for an interview, was offered the job, and accepted. In spring 2008, I was off to California. Bea would soon follow, we hoped.

Like TNC, PRBO was a nonprofit conservation organization. But the similarities ended there. TNC was one of the world's largest conservation organizations; PRBO was more than an order of magnitude smaller. TNC had an annual budget of over $1 billion; PRBO's budget was less than $10 million. TNC had large departments at the worldwide office in Arlington devoted to the legal support for real estate transactions, fundraising, and international conservation policy. PRBO had no staff dedicated to legal affairs and only three people directly involved in philanthropy. And although some of the PRBO scientists worked internationally (on penguins in Antarctica, for example), there was little direct involvement with international organizations or policy.

The most telling differences, however, were the location and staff. PRBO was based in Petaluma, an hour north of San Francisco. Instead of taking the Metro or driving to my office in an eight-story building in the heart of the Washington, DC, metropolitan area (population over six million), I could bike to work in a small two-story building in Petaluma (population less than 60,000). PRBO's office was adjacent to a large wetland bordering the Petaluma River, so I could walk out the door and immediately begin enjoying avocets, black-necked stilts, marsh wrens, and flocks of white pelicans. And most of PRBO's staff were scientists who were conducting hands-on scientific studies to support the efforts of state or federal agencies and other organizations. They were actually *doing* science. Many of them had been working together at PRBO for years or decades, so there was a collegial atmosphere. It was easy to find someone to talk science with or share ideas. And their work centered almost entirely on birds.

My role as chief scientist at PRBO evolved over time. Although I supervised several of the senior scientists, there were few administrative responsibilities. I met with agencies, public organizations, potential donors, and academic groups to learn about their conservation concerns and promote PRBO's science. I worked with PRBO's small board to explain the importance of strong science to effective conservation. And as I had done at TNC, I mentored the science staff in designing research projects, preparing presentations for conferences or papers for publication, and helping them develop their own career pathways. But I could also be an active participant in research projects. We brainstormed ideas and approaches, analyzed and interpreted data, and discussed how best to present our findings to user groups, other scientists, and the public. Three of these projects illustrate the sort of direct involvement in conservation science that I had missed while at TNC. My conservation pathway was becoming more sharply defined as a conservation *science* pathway.

One project took me back to seabirds. Marine Protected Areas (MPAs) are one of the most effective measures that can be taken to conserve marine ecosystems and the fish, seabirds, marine mammals, and other biota they support. PRBO had an active marine conservation program, which included conducting at-sea shipboard surveys of seabirds, marine mammals, and their food resources. Nadav Nur, Jaime Jahncke, and several other PRBO scientists got to wondering whether the information from such surveys could be used to evaluate the effectiveness of MPAs on the West Coast—specifically, whether "hotspots" where multiple seabird species aggregated in areas of high food availability were protected within existing MPAs. Using data from surveys and bathymetric and oceanographic data, we predicted the distribution and abundance of sixteen seabird species in the California Current area. We then combined the results across species to determine the locations of seabird hotspots. Many of the hotspots did indeed occur in existing MPAs, confirming their conservation value. But our analysis also identified several areas that were not included in any MPA. We suggested that the seabirds were telling us that these areas should be prioritized for future protection.

We published our results in the scientific literature and shared our findings with agencies on the West Coast.[1] Our efforts to promote our work to influence policy, however, ended there. Unlike TNC, PRBO did not have a large government relations program dedicated to translating conservation practice into policy.

~~~~~~

The effects of climate change on bird populations and communities and their habitats figured importantly in PRBO's activities, so many of the projects on which I collaborated involved climate change. When I arrived at PRBO, Diana Stralberg, Dennis Jongsomjit, and several others were initiating a new project dealing with birds and climate change. They were asking three questions: How might the distributions of bird species in California be affected by future climate change? Would suites of species be affected in the same way, or would the species respond to climate change independently of one another? How might the changes affect the overall composition of bird assemblages or communities? These questions got to the heart of relating potential climate changes to bird populations. The answers could affect how protected areas are managed or where organizations like TNC target lands for acquisition. I was asked to join in, which I did. It was a long-awaited opportunity to be directly engaged in applying science to questions about climate change and bird conservation.

Short of reading tea leaves or astrological signs, models are the only way to gain a glimpse into what the future may hold. Building on models that correlated current distributions of bird species with data on the current distribution of major vegetation types, we projected future bird distributions based on how the future climate (derived from global-circulation models) would affect habitat, and thus the occurrence of a species. We then calculated the change between the current distribution of each of sixty focal bird species and their projected future distributions in the year 2070.

Most species were projected to contract their distribution with future changes in climate and habitat.[2] As the species responded to the changing conditions, their distributions would shift independently of one another. As a result, some species that currently do not co-occur might be found together in the future, and vice versa. For example, wrentits and rufous-crowned sparrows currently co-occur only in limited areas of scrub habitat in central California. The model projections suggested that these species will be found together far more frequently in the future as sparrows move northward and wrentits shift toward coastal regions and upslope in the Sierra foothills. As species respond individualistically to future climate change, community composition will be reshuffled, creating novel or "no-analog" assemblages of species that have not occurred together before. Overall, the models projected that more than half of the state might contain no-analog bird communities by 2070. These areas would present birds, as well as managers and conservationists, with unprecedented challenges. Widespread reshufflings of communities and the emergence of novel ecosystems may in fact be one of the most profound ecological consequences of climate change. The future, it seems, will be very different from the past or present.

～～

Climate change is not the only environmental disruption confronting bird populations and communities. Changes in human land use may be a more immediate threat. Taking this cue, several of us realized that we might be able to build on our climate change modeling to consider the additional effects of future land-use change on California birds.

"Land use" encompasses a wide variety of actions. To simplify our analysis, we used housing density (development) as a surrogate measure of land use. Housing density in urban and suburban areas in

California is projected to nearly double by 2070, while exurban development is expected to triple. We asked how changes in housing density might alter our projections of climate-induced changes in habitats and bird distributions.[3]

The distribution of nearly two-thirds of the species we considered was projected to contract with climate change. For some of these species, changes due to development could match or exceed the climate-related losses. The distribution of other species was expected to expand with climate change, but for nearly half of these species the addition of development to the models reduced or completely counteracted the climate-related distributional gains. For example, California gnatcatchers are listed as a threatened species under the U.S. Endangered Species Act, which brings conservation priorities into conflict with future development plans. Climate models alone suggested that gnatcatchers would expand their distribution northward as their coastal scrub habitat expands. Yet much of the gnatcatcher's current distribution lies in the rapidly growing Los Angeles-San Diego corridor. The northward expansion would likely be stymied and fragmented by development.[4] Clearly, considering only climate change or development by itself could produce an incomplete picture of the management challenges. For example, species associated with coniferous forests might benefit from management focused on climate-related impacts, while managing housing densities might be critical for chaparral-scrub species.

The initial focus of this modeling exercise was on a question of basic science: How would the potential impacts of changing climate and development affect bird distributions in California? With support from the Legacy Resource Management Program of the Department of Defense (DoD), we extended our modeling to address a more applied question: How might the changes in bird populations and communities affect environmental management on military installations or lands administered by the DoD in California?

Our modeling analyses projected that future losses in bird-species richness due to climate change would be greater on DoD lands than the average for the state as a whole or on areas administered by other federal agencies. However, the effects on bird distributions of housing developments immediately adjacent to DoD lands were often greater than those of climate change alone: of twenty-three installations showing climate-induced reductions in bird-species occurrences, ten had even larger reductions due to housing changes.

We distributed our report to environmental resource managers in the military agencies, summarized our results in internal DoD newsletters, and conducted briefings in response to inquiries from personnel on several DoD bases. As with the seabird/MPA project, however, applying the results of our analyses to carry the project forward to directly influence management practices or policies was beyond the organizational capacity of PRBO.

These exercises all relied heavily on model projections. How much faith should managers and conservationists have in such projections? Should modeling results be used to guide conservation actions, or are they only illusions? Models are always plagued by uncertainties, especially when climate change is involved.[5] Yet it may often be necessary to move ahead despite the uncertainty of models. Striving for a high level of scientific certainty in projecting future conditions is unrealistic; rather than waiting until we have better models (or the future has become the present), a glimpse of probable futures may be good enough to initiate management or conservation actions.[6]

I spent six years at TNC and four more at PRBO. My decision to embark on the conservation science pathway came from a convergence of *people*, *interests*, and *opportunities*, rather than from thoughtful planning. Although I was comfortable in my academic position and enjoyed my work in landscape ecology, I couldn't let the opportunity to join a major conservation organization in an influential role pass. So I immediately jumped into a total immersion in conservation. I wasn't just doing research with conservation implications or writing about conservation issues any longer; I was submerging myself in the daily doings of conservation. And I could scarcely have picked two better organizations to show me how conservation is done and how science can strengthen the process than TNC and PRBO.

TNC covered virtually all aspects of conservation practice. There was science, of course (that's what brought me to TNC), but programs for fundraising, government relations, policy, legal affairs, communications, public relations, and other activities were also well-staffed. At the state level, chapters worked closely with state officials and landowners. I remember hearing a TNC preserve steward in Pennsylvania describe how he had nurtured a local farmer to convince him to grant TNC a conservation easement for a stretch of stream on his property. He got together with the landowner often, bringing beers to share while they sat on his porch and talked of baseball and many things other than conservation. After several years, he established trust, and the farmer granted the easement. Staff in TNC's international programs often made similar efforts, tuning their approaches to local traditions, cultures, and politics. Science didn't always figure into such efforts, but it was an important part of the planning process used to identify places, such as the Pennsylvania stream, that merited conservation. TNC implemented conservation actions and emphasized outcomes.

PRBO was much smaller than TNC and, consequently, differed in its conservation focus. Scientists in PRBO undertook the in-depth scientific studies required to understand a conservation challenge and inform conservation practice and management. Although PRBO made efforts to reach out to landowners, agency personnel, and legislators (mostly in California), it was up to others to take appropriate actions. In some cases, PRBO scientists asked the questions that they thought needed to be answered to address a conservation problem. In other cases, resource managers or agency personnel asked the questions, hoping that PRBO scientists could come up with a science-based answer. In either case, the answers provided information to guide actions—where to put MPAs to protect seabirds, how climate change might affect bird community composition, and where future housing development might override the effects of climate change. Others would undertake the conservation actions.

TNC gave me an appreciation of the many interacting components of conservation writ large, while PRBO enabled me to actually do conservation science. It was an effective combination. Conservation was my job, but it was fueled by my interests. Although I was still following a landscape ecology pathway, it was now interwoven with the conservation pathway, reinforced by my abiding interest in birds.

Eventually, however, my active involvement with conservation organizations reached an end. I left PRBO in 2012 when I officially retired. Bea had taken a position as program leader at the Forestry Sciences Laboratory of the U.S. Forest Service in Corvallis, Oregon. It was well past time to reunite, so I joined her there. Bea kept on in her position for several years, and then spent two years as director of the newly instituted Northwest Climate Hub of the U.S. Department of Agriculture (still in Corvallis) before she also retired in 2018. We never thought when we left Corvallis in 1986 that we'd end up back there again, but there we were. Although I continued my journeys along several pathways, I now did so by writing about the things that interested me: birds, ecology, landscapes, conservation, and whatever else captured my attention.

25

WRITING INTO RETIREMENT

What do scientists do when they retire? Many forsake the professional activities that have defined their lives for decades to pursue entirely different dreams. They cancel their memberships in professional societies, sell or donate their scientific books, no longer write scientific papers or grant proposals, and sever their ties with academia and former colleagues and collaborators. They are no longer interested in what once excited them. If they are in academia, they may have been worn down by the demands of classes, committees, and campus politics and become discouraged by the prospects of getting grants to continue their research and support graduate students. They go fly-fishing, play golf, do carpentry, take up cooking, travel to exotic places, or otherwise immerse themselves in what may previously have been their hobbies. They have abandoned their former pathways.

Others see their current pathways extending in front of them, beckoning them to explore further. They can't envision leaving behind what had interested and excited them for an entire career. For me, momentum and motivation, fueled by my interests and enabled by opportunities, kept me moving along the pathways I had been following. "Retirement" meant that I could now do what I wanted to do with no job obligations. And what I wanted to do most was to write about interesting things.

I had long believed that, while it might be unrealistic to expect scientific writing to be compelling prose, at least it could be clear and grammatically correct. I had an opportunity to put this belief into practice when I edited an ornithological journal, *The Auk*.[1] In the mid-1970s, I was fulfilling my professional obligation to the ornithological pathway that I was on by serving as treasurer of the American Ornithologists' Union (AOU). The governing council had decided it was time to replace the longtime editor of the journal and I had expressed some interest. During a break, when the council was debating what to do, I was approached by Ernst Mayr. Mayr was a former president of the AOU and arguably the world's preeminent evolutionary biologist—someone who commanded immediate respect. He asked why I was interested in the position. I mumbled something about the opportunity to advance ornithology as a science and my belief in the value of good writing. This was apparently enough to convince him to advocate for me in the council's deliberations, and I was appointed editor.

I ended up editing *The Auk* for eight years (1976–1984). Mostly this involved shepherding manuscript submissions through the review process and making decisions about what and what not to publish. Sometimes these decisions were difficult. On more than one occasion, I had to reject a submission from an iconic leader in ornithology, someone I'd grown up admiring. Now and then I overrode reviewers' recommendations and published a paper that was clearly on the edge of conventional thinking. Sometimes these papers helped to ignite new research; more often they became neglected curiosities. Occasionally I had to deal with ethical issues, such as attempts to publish the same material multiple times in different journals (an unfortunate manifestation of the growing "publish or perish" pressures faced by many scientists).

Editing *The Auk* involved working to make other people's writing better. But I'd also been doing more and more writing of my own.

Much of this was in papers or book chapters on weighty scientific topics. But that was scientific writing, and although I tried to make it more readable than the standard scientific fare, it still had to adhere to a particular style. Eventually, however, my undergraduate interests in the humanities and literature, long suppressed, began to emerge full force. I wanted to do something that would express my opinions and thoughts in literary prose rather than scientificese. I could write essays.

EPISODE: THE ESSAYIST

I first started writing essays when I was editing *The Auk*. I used my editorial prerogative to offer my opinions on things that I had seen in manuscripts which were submitted for publication—a tendency to avoid skepticism, an overreliance on word processing software rather than actual writing skills, and the like. But when I quit editing the journal I lost my platform, and for some time I wrote essays only occasionally. Then in 2007, Alan Crowden invited me to contribute essays to the *Bulletin of the British Ecological Society*. Alan was editing the *Bulletin*, which appeared quarterly, and he thought that essays might be a nice addition to the usual fare of standard business and reports.

Once again it was *people*—in this case Alan—who played the critical role in setting me off in this direction. I had gotten to know Alan some years before in his capacity with Cambridge University Press. Alan's responsibilities there included soliciting potential book projects from ecologists and like scientists, and he had asked me several times to comment on project proposals. Now he was asking me to send him essays for the *Bulletin* "now and then." It turned out to be four times a year for twelve years.

The idea appealed to me. The *Bulletin* was distributed to all members of the British Ecological Society, so the readership of my contributions would primarily be other ecologists, most of them in the UK and Europe. It would give me an opportunity to vent my opinions on the topics of my choosing. This is something not normally allowed in scientific papers, which are supposed to hew to objectivity. By this stage of my career, I had many opinions waiting to be vented. I might be able to use the essays to influence how other ecologists were thinking. The essays would also allow me to try my hand in writing actual prose, in which words and how they are used and strung together become an art. I wasn't sure I could do this, having been so thoroughly immersed in scientific writing for so many years. But it would be fun to try.

It wasn't as easy as I had thought. Because Alan told me I could write on any subject, I had to come up with topics.[2] This was often spur-of-the-moment as one of Alan's "deadlines" approached. My choice of a topic was usually triggered by a news item, a recent event, a book I'd read, or simply something that occurred to me that I thought would be fun to write about. Sometimes it was classical literature that provided a lead-in—Dickens's "It was the best of times, it was the worst of times," or Shelley's poem *Ozymandias,* for example. C. P. Snow's *The Two Cultures* provided the impetus for one essay. I wrote about scientific or ecological topics such as uncertainty, paradigms, ecosystem services, and the value of nature, but I also wrote about the green (i.e., environmentally responsible) house that Bea and I built and the conventions against using footnotes in scientific writing. The writing entailed incessant revision and tuning to find just the right way of expressing an idea. I thoroughly enjoyed it.

It turned out that Alan had also invited Richard Hobbs to contribute essays with an antipodean theme ("From Our Southern Correspondent"). Richard was an ecologist at the University of Western Australia in Perth, where he led a large group of students and postdocs working on ecological restoration. We had previously collaborated on scientific publications and symposia at scientific conferences. Richard had arranged for me to be awarded a Winthrop Research Professorship at the university, which supported several visits to Perth during which Richard and I could collaborate on whatever we wished. Richard and I spent many hours pondering how and why Australian and North American ecologists and ecology differed. We often adjourned to Little Creatures brewery in Fremantle to conjure up joint essays and conspire about how to convince Alan that his "deadlines" meant nothing to us (something he'd already discovered). We explored various facets of the Australia–North America differences in ecologies, and even wrote one essay as a dialog (something you could *never* do in a proper scientific publication).[3]

I was proud of the essays. After I had been writing them for years, I began to think that they might appeal to a wider audience. I approached a publisher (John Wiley & Sons) about the possibility of putting together a collection of essays. To provide some glue to bind the thirty-eight disparate essays together, I wrote a series of short reflections on various topics. The book, *Ecological Challenges and Conservation Conundrums*, was published in 2016.[4] I continued to write additional essays until 2018, when Alan retired from editing the *Bulletin* and Richard and I, having lost our muse, retired from writing essays.

My writing dealt with a wide variety of topics and issues, some more important than others. Perhaps the most important, in terms of its implications for conservation, began in a collaboration with Mike Scott. Mike had been one of my first graduate students and we had worked together in the early 1970s when we modeled seabird energetics (see chapters 9 and 19). Mike had gone on to a career with the U.S. Fish & Wildlife Service focusing on the conservation of endangered species. Later he became leader of the USGS Cooperative Fish and Wildlife Unit at the University of Idaho, where he joined forces with Dale Goble, an environmental law professor. Mike, Dale, and I began discussing the problem of *conservation-reliant species*. Most efforts to conserve threatened species presumed that if appropriate management was implemented the species would recover and no longer require intensive management. We proposed instead that the factors that threatened species and led to their decline or imperilment in the first place were often likely to persist—fragmented habitat is not easily made whole and invasive species are not easily eradicated. Management might reduce the threats and allow a species to stabilize or recover, but if the management efforts stopped the threats would reemerge and the species would slide back into imperilment. The species would be reliant on continuing, long-term management. The costs of this ongoing management were not usually considered in conservation plans.

We published several papers calling attention to the problem.[5] Our arguments prompted many responses, most of them negative and some of them outraged. Some thought we were casting doubt on the value of conservation, painting full recovery as an unattainable goal. As it became more widely known, the notion of conservation reliance generated even more backlash. We decided that the issue required a full, book-length treatment.[6] Joined by Bea, we pulled together examples of conservation-reliant species from throughout the world, assessed threat factors and the effectiveness of various mitigation practices, considered the role of laws and regulations, and addressed the challenge of implementing effective management in a rapidly changing world. The implications of conservation reliance were even more profound than we had originally thought. We were calling attention to a basic conundrum in conservation. Because the threats to many imperiled species cannot be eliminated, management (and its costs) might need to continue, perhaps for a very long time. But resources necessary to implement management for all at-risk species are not likely to be available, so allocation of resources among species will need to be prioritized. This will leave some species to deal with threats on their own. Some will fail.

This was a hard conclusion for ardent conservationists to accept. "Saving all species" had become a mantra of conservation, and our arguments clearly implied that this was not possible. We were suggesting that not all imperiled species can be saved and choices must be made—conservation triage. Again, I found myself arguing against widely held beliefs (or hopes).

Scientists must write. Publications are one of the most important metrics of success in science. Scientific papers are how you inform your colleagues about what you've done and what you've found out. It is especially important to publish often as you are beginning your career, to establish a track record and recognition among your peers.[7] Because papers that follow a standard format are more likely to be accepted for publication, there are strong pressures to write in a prescribed style. Some journals even require papers to follow a set style, so that everything reads the same way. By the time you've become well-established, you may have become comfortable writing scientifically. But other, like-minded scientists may not be your only audience. To reach a broader readership you will need to abandon the formulaic, scientific style that has worked for you before, and turn to writing readable and interesting prose. This can be hard. The best way to do it, I think, is to continue reading nonscientific literature and appreciate how good writing makes you feel. There is still much to be learned from reading the classics.

My own desire to write probably reflected my family history, my early interests in the humanities, and the considerable influence of people such as Frances Dunham (my high school English teacher) and George Sutton, people who impressed upon me the value and joy of good writing. As I moved into retirement, writing came to dominate my scientific activities. But writing was not a new or different pathway. Rather, it was a way of consolidating and sharpening my thinking and expressing my ideas and opinions no matter which pathway I was following. The papers, book chapters, and essays I wrote about competition, ecological communities, patchiness, scale, and landscape ecology helped form the conceptual foundations of several disciplines, and my writing about conservation drew attention to the vexing issue of conservation reliance. The pen is powerful, even in science.

CONCLUSIONS

What Are the Lessons for Today's Aspiring Ecologists?

I n the previous chapters, I've taken you on my journey of becoming an ecologist. It's been an evolving web of pathways, reflecting the times, my shifting interests, and the sudden appearance of opportunities I didn't anticipate. Over the decades I changed—but ecology also changed. The power of the competition paradigm that dominated community ecology when I was growing up as an ecologist faded as anomalies accumulated and the variability of nature became more widely appreciated. Ecologists became more interested in the societal implications of their work. Conservation, with its roots in ecology, gained momentum as awareness of the perilous state of the Earth's biodiversity grew. New technologies enabled ecologists to probe nature's secrets in increasing detail with greater sophistication. And ecology will continue to change as new technologies and artificial intelligence lead to new questions.

The pathways I followed were just one route of many to becoming an ecologist. Someone interested in ecosystems might follow pathways into modeling or nitrogen cycling, and the pathways of a population ecologist might be different still. Until I distracted him,

Rex Cates's pathway emphasized plant secondary chemistry. There are many ways to explore the relationships between organisms and their environments, all of which can be labeled "ecology."

I've argued throughout this book that, regardless of the particular route you take to become an ecologist or how "ecology" is defined, the pathways are determined by three driving forces: your *interests*, your *motivation*, and the *opportunities* that come your way. Interests prompt questions, which are the stuff of which science is built. Like the baby elephant in Kipling's *The Elephant's Child*, an ecologist is full of "'satiable curtiosity" which drives him or her to ask, "a new fine question that [he] had never asked before."[1] Although interests often first appear in childhood, they must be nurtured and directed to lead to a career in science. Mentoring is critical. Mentoring can come from many sources—parents, relatives, and teachers, to be sure, but also people like camp counselors or members of a local bird club. These people can keep interests alive, pushing you to look more deeply to ask those "new fine questions." As your career develops, interests become better defined. Paradigms can focus interests, but they can also stifle them.

Interests alone are not sufficient. A person may have plenty of interests, but few of them develop into a career pathway. Interests must be fueled by motivation. Someone may be motivated to become an ecologist because they want to address environmental concerns, want to protect threatened species, love birds or cacti, or simply want to find things out. Motivation is the passion that leads you to follow interests, embark on a pathway, and then keep going. People also play an important role in reinforcing motivation, especially when you are just getting started or are discouraged when things aren't working out as planned.

Interest and motivation lead you to choose a pathway to follow. Opportunity determines whether a pathway that aligns with your interests and motivation will be available to you. Opportunities often appear suddenly when they are least expected. How you encounter and respond to opportunities can determine the trajectory of your career pathways. Unfortunately, opportunities are not equally available to everyone when they do appear. Discrimination on the basis of gender, race, ethnicity, religion, or culture still restricts opportunities for some people more than others, although not to the degree it did when I started down an ecology pathway. The opportunity to become an ecologist should be available to everyone, not only because that's what's right but because the varied perspectives and insights enrich ecology (and ecologists).

I started on my pathway to becoming an ecologist over sixty years ago. Times have changed. The experiences that help spark the interests of budding ecologists have become more restricted. Parents no longer feel safe letting their children encounter nature unfettered and unsupervised, as I did. Students may be more likely to learn about nature virtually than directly, and the nature that they do experience is often only a faint picture of what it once was. Perhaps I'm just an old curmudgeon complaining, but I worry that spending time staring at a computer screen will not produce the insights and inspiration you can get from sitting on a log in the woods and observing nature.[2]

Training to become an ecologist has also become more demanding. As ecology has become more rigorous and scientific, academic curricula have been loaded with required courses—not just several levels of ecology but also some combination of mathematics, statistics, biochemistry, plant identification, vertebrate biology, computer modeling, and (still in some places) a foreign language or two.[3] And it now requires additional training to use the technology and tools available to ecologists. Hopscotching through an eclectic variety of coursework, as I did, is no longer feasible if you want to complete a degree program in a reasonable amount of time. There is too much to learn. Aspiring ecologists no longer have the time to take courses in Shakespeare, philosophy, or art. I would argue that they are poorer for it.

You must also get a job and join the scientific culture. Once you have started down a particular pathway in ecology the pressures to stay on that pathway can be formidable. To advance, you must obtain research grants and publish several peer-reviewed papers a year. It's no longer possible to obtain grant funding for general, largely observational projects that address no explicit hypotheses, as I did when I was starting out. There is a premium on posing questions that can be answered quickly and are of interest to your peers. Pursuing your interests wherever they might lead becomes an unaffordable luxury. Better, perhaps, to stay on the straight and narrow.

But the variability of natural systems means that the longer you study a system the greater the likelihood that you will encounter anomalies, especially as the accelerating pace of environmental change makes extreme events more frequent and more extreme. Anomalous observations can make it difficult to generate ongoing funding and lead you to question whether continuing on a pathway is really such a good idea. I probably would have stayed on some pathways longer if my results had matched expectations more often.

You don't take a long journey along several related career pathways, as I have done, without learning something along the way. What have I learned? Looking back, I can see several lessons. I didn't recognize these at the time, yet somehow I managed to have a successful career in several disciplines: ornithology, community ecology, landscape ecology, and conservation. Now, with the benefit of hindsight I can offer some advice that may be useful if you are starting out or already following a pathway to become an ecologist. As I said earlier, "becoming an ecologist" is an ongoing adventure.

Here, then, is some of what I've learned in my journey to become an ecologist.

Let your interests be your guide. Study what excites you about nature. To be a good ecologist you must enjoy it. Ask yourself why you would want to become an ecologist. What is it about nature that arouses your passion?

Be aware that how your questions are posed and your studies are designed affects their outcomes and your interpretations of the results. This is one of the emerging lessons of the *Exxon Valdez* investigations (see chapter 21): casting the investigations at the outset with an initial assumption of a negative impact or a null expectation of no impact influenced all that followed.

Pay attention to what questions interest your colleagues and peers. These are the people who will review your funding proposals and scientific papers. They may determine whether you will be able to join the community of scientists and advance along a pathway. They may often be conducting their studies under the umbrella of a paradigm.

Be true to your observations and data. If you have designed and conducted your studies carefully and framed your questions so they can be answered, your results should be trusted. If they do not match expectations, your study is not a failure; you may have uncovered some of nature's anomalies. If your findings lead you to challenge a paradigm or a popular theory, do so, but be thoughtful and diplomatic.

Don't ignore or disregard anomalous observations. Anomalous results are only anomalies in the context of an accepted theory or paradigm. They may tell you more about how nature works than would confirmatory results. Anomalous results should prompt you to consider alternative explanations, which may lead you to ask new questions or embark on a new pathway.

Don't let yourself become a slave to tools or technology. The questions you ask as an ecologist should be about how nature works, not about how you can best use a particular tool. Use new tools if they enable you to probe more deeply into a problem, not simply because they are novel and trendy.

Seek out collaborators when they can help you address questions that otherwise would be beyond your reach. Just be sure that your personalities don't clash and that you enjoy the collaboration.

Diversify your experiences. Diverse experience is helpful when it comes to getting a job, and it can bring fresh insights and approaches to your studies. Take advantage of opportunities to participate in research projects and take nonscience courses in college. Read good literature and travel widely when you can. Avoid overspecialization.

Consider the confounding effects of environmental variation and scale when planning a study. Think about whether your question can be addressed in a short-term study in a single place or requires a longer investigation in several locations. Ask whether the scale of your study is appropriate for the organisms, phenomena, and questions you want to investigate.

Be alert to the societal relevance of your work. Many basic-science investigations in ecology may have broader societal relevance if the questions are framed in a different way. Casting your work to address societal issues may open new sources of funding and lead you to new pathways. Besides, it is satisfying to think that your work might be useful and have broad applications.

Beware of advocacy. Remember that as an ecologist you are an advocate for the integrity of the scientific process. Guard against the temptation to use the results of your studies to advance a particular agenda, particularly if there is pressure to tune your results to support a foregone conclusion.

Be thoughtful about switching pathways. Don't let inertia bind you to your current pathway if an opportunity suddenly appears to embark in a new direction. Changing pathways can be invigorating, particularly if your interest in your current pathway is flagging. But recognize that changing pathways has costs, especially if it entails a change in disciplines. You will need to develop a new network of colleagues, attend different conferences, submit your papers to different journals, and appeal to different sources of funding. All of this can lead to a loss in momentum and stifle your professional progress. Look before you leap.

Don't hesitate to detour onto side pathways. You never know where they might lead. And even if they don't lead to a change in pathways, they may be productive if they interest you. But let your interests be your guide.

And finally . . .

Take pleasure in finding things out. That's what science is all about. That's why you become an ecologist.

APPENDIX

SCIENTIFIC NAMES OF SPECIES MENTIONED IN TEXT

Blackbird, Red-winged	*Agelaius phonecius*
Bufflehead	*Bucephala albeola*
Bunting, Lark	*Calamospiza melanocorys*
Cormorant, Brandt's	*Phalacrocorax penicillatus*
Cowbird, Brown-headed	*Molothrus ater*
Dickcissel	*Spiza americana*
Duck, Harlequin	*Histrionicus histrionicus*
Frog, Red-legged	*Rana aurora*
Frog, Cascade	*Rana cascadae*
Galah	*Eolophus roseicapilla*
Gnatcatcher, California	*Polioptila californica*
Goldfinch, Lesser	*Spinus psaltria*
Grackle, Common	*Quiscalus quiscula*
Grebe, Horned	*Podiceps auritus*
Guillemot, Pigeon	*Cepphus columba*
Gull, Glaucous-winged	*Larus glaucescens*
Gull, Western	*Larus occidentalis*
Kangaroo, Red	*Macropus rufus*
Lark, Horned	*Eremophila alpestris*

Longspur, McCowan's	*Rhynchophanes mccownii*
Lorikeet, Rainbow	*Trichoglossus moluccanus*
Lyrebird, Superb	*Menura novaehollandiae*
Lyrebird, Albert's	*Menura alberti*
Magpie, Australian	*Gymnorhina tibicen*
Meadowlark, Eastern	*Sturnella magna*
Meadowlark, Western	*Sturnella neglecta*
Murre, Common	*Uria aalge*
Oriole, Orchard	*Icterus spurius*
Owl, Barred	*Strix varia*
Owl, Spotted	*Strix occidentalis*
Peeper, Spring	*Pseudacris crucifer*
Phalarope, Red-necked	*Phalaropus lobatus*
Pipit, Richard's	*Anthus richardi*
Sandpiper, Pectoral	*Calidris melanotos*
Shearwater, Sooty	*Ardenna grisea*
Songlark, Brown	*Cincloramphus cruralis*
Sparrow, Black-throated	*Amphispiza bilineata*
Sparrow, Brewer's	*Spizella breweri*
Sparrow, European Tree	*Passer montanus*
Sparrow, Grasshopper	*Ammodramus savannarum*
Sparrow, House	*Passer domesticus*
Sparrow, Rufous-crowned	*Aimophila ruficeps*
Sparrow, Sage	*Artemisiospiza nevadensis*
Sparrow, Savannah	*Passerculus sandwichensis*
Sparrow, Song	*Melospiza melodia*
Storm petrel, Leach's	*Hydrobates leucorhous*
Thrasher, Sage	*Oreoscoptes montanus*
Tit, Blue	*Cyanistes caeruleus*
Tit, Great	*Parus major*
Vireo, Black-capped	*Vireo atricapilla*
Warbler, Kirtland's	*Setophaga kirtlandii*
Whiteface, Southern	*Aphelocephala leucopsis*
Wrentit	*Chamaea fasciata*

NOTES

INTRODUCTION: CHARTING PATHWAYS

1. Malcolm L. Hunter, David B. Lindenmayer, and Aram J. K. Calhoun, *Saving the Earth as a Career: Advice on Becoming a Conservation Professional, Second Edition* (Chichester, UK: John Wiley & Sons, 2016); Ashley Juavinett, *So You Want to Be a Neuroscientist?* (New York: Columbia University Press, 2020); and Roy Plotnick, *Explorers of Deep Time: Paleontologists and the History of Life* (New York: Columbia University Press, 2022), respectively.

2. To enhance the readability of my narrative I have omitted most of the sciency details. I provide literature citations in endnotes and the bibliography for those who would like to dig deeper.

3. One can buy Ecology bread or drink suds from an Ecology beer glass, for example, and the term "ecosystem" has been appropriated to refer to any complex, interconnected system (e.g., the "business ecosystem").

4. Recognition of the relevance of ecology to human affairs is not new. In the ecology textbook that was most widely used when I was starting out, Eugene P. Odum defined ecology as "the study of the structure and function of nature" but added, parenthetically, "it being understood that mankind is a part of nature." *Fundamentals of Ecology*, 2nd ed. (Philadelphia: W. B. Saunders, 1959), 4.

5. It is this element of activism that led some to label ecology (rather proudly) as "the subversive science." Paul B. Sears, "Ecology—a Subversive Subject,"

BioScience 14, no. 7 (July 1964): 11–13; Garrett Hardin, "Human Ecology: The Subversive, Conservative Science," *American Zoologist* 25 (1985): 469–76.

6. I am reminded of a quote attributed to Yogi Berra: "When you come to a fork in the road, take it." Berra also said that "I never said most of the things I said."

7. Thomas S. Kuhn, *The Structure of Scientific Revolutions*, 2nd ed. (Chicago: University of Chicago Press, 1970); see John A. Wiens, *Ecological Challenges and Conservation Conundrums. Essays and Reflections for a Changing World* (Chichester, UK: John Wiley & Sons, 2016), 18–32.

8. Kuhn defined normal science as "research firmly based upon one or more past scientific achievements . . . that some particular scientific community acknowledges for a time as supplying the foundation for its further practice." Kuhn, *The Structure of Scientific Revolutions*, 10.

9. See, for example, Imre Lakatos and Alan Musgrave, eds., *Criticism and the Growth of Knowledge* (London: Cambridge University Press, 1970), 1–278; and Thomas Nickles, ed., *Thomas Kuhn* (Cambridge: Cambridge University Press, 2003), 1–284.

10. See the exchange between Thomas W. Schoener, ["Reply to Wiens 1983"], *American Scientist* 71 (1983): 235, and John A. Wiens, "Interspecific Competition," *American Scientist* 71 (1983): 234–35; and the treatment of the issue by Kim Cuddington and Beatrix Beisner, eds., *Ecological Paradigms Lost: Routes of Theory Change* (Burlington, MA: Elsevier, 2005).

11. And interpret them, and even write your paper if you are so inclined. But then you would no longer be a scientist, but a nursemaid to a bot.

1. IN THE BEGINNING: EMERGING OF INTERESTS

1. The title of Richard Feynman's collection of short essays. Richard P. Feynman, *The Pleasure of Finding Things Out* (Cambridge, MA: Perseus Books, 1999).

2. E. O. Wilson, *Naturalist* (Washington, DC: Island Press, 1994).

3. Richard Louv, *Last Child in the Woods: Saving our Children from Nature-Deficit Disorder* (London: Atlantic Books, 2005).

4. John Wiens and Charles Krebs, "Last Ecologist in the Woods?," *Bulletin of the British Ecological Society* 48, no. 2 (2017): 47–49.

5. Aldo Leopold, *Round River*, ed. Luna B. Leopold (New York: Oxford University Press, 1953), 165.

6. A portrait of a purple finch that Sutton painted for my grandmother is featured on the cover of Jerome Jackson's biography of Sutton. Jerome A.

Jackson, *George Miksch Sutton: Artist, Scientist, and Teacher* (Norman: University of Oklahoma Press, 2007).

7. Margaret Morse Nice, "Studies in the Life History of the Song Sparrow, Volume 1, A Population Study of the Song Sparrow," *Transactions of the Linnean Society of New York* 4 (1937): 1–247; see Marilyn Bailey Ogilvie, *For the Birds: American Ornithologist Margaret Morse Nice* (Norman: University of Oklahoma Press, 2018) for a recent biography of Margaret Nice, a truly amazing woman who was years ahead of her time.

8. George Miksch Sutton, letter to author, June 4, 1956.

2. MOLDING OF INTERESTS

1. George Miksch Sutton, letter to author, August 26, 1959.

2. Jerome A. Jackson, *George Miksch Sutton: Artist, Scientist, and Teacher* (Norman: University of Oklahoma Press, 2007), 114.

3. George Miksch Sutton, *Oklahoma Birds: Their Ecology and Distribution, with Comments on the Avifauna of the Southern Great Plains* (Norman: University of Oklahoma Press, 1967).

4. eBird, https://ebird.org.

5. J. Michael Scott et al., *Shepherding Nature: The Challenge of Conservation Reliance* (Cambridge: Cambridge University Press, 2020).

6. Fledging success was low because many nests lost eggs or chicks to snake predation.

7. John A. Wiens, "Aspects of Cowbird Parasitism in Southern Oklahoma," *Wilson Bulletin* 75 (1963): 130–39.

8. A decade later, Kendeigh and I would find ourselves collaborating on studies of granivorous birds and publishing together. I am sure he had long since forgotten about our interaction, and I never reminded him.

9. Wiens, "Aspects of Cowbird Parasitism," 130.

3. STARTING ON THE PATHWAY TO BECOMING A SCIENTIST

1. Paul L. Errington, *Of Men and Marshes* (Ames: Iowa State University Press, 1957).

2. The issue of how the act of observing something affects what is being observed, with the result that the observations are no longer independent of the observer, is an expression of the Heisenberg Uncertainty Principle

of quantum mechanics; it is a pervasive problem in studies of animal behavior and ecology.

3. John A. Wiens, "Behavioral Interactions of Red-Winged Blackbirds and Common Grackles on a Common Breeding Ground," *The Auk* 82 (1965): 356–74.

4. DEFINING A PATHWAY

1. Robert H. MacArthur, "Population Ecology of Some Warblers of Northeastern Coniferous Forests," *Ecology* 39 (1958): 599–619.

2. Martin L. Cody, letter to author, January 3, 1965.

3. John A. Wiens, "An Approach to the Study of Ecological Relationships Among Grassland Birds," *Ornithological Monographs* 8 (1969): 89.

4. John A. Wiens, "Interterritorial Habitat Variation in Grasshopper and Savannah Sparrows," *Ecology* 54 (1973): 877–84.

5. V. C. Wynne-Edwards, *Animal Dispersion in Relation to Social Behaviour* (New York: Hafner, 1962), 1–630.

6. Thomas Robert Malthus, *An Essay on the Principle of Population* (London: J. Johnson, 1798).

7. John A. Wiens, "On Group Selection and Wynne-Edwards' Hypothesis," *American Scientist* 54 (1966): 273–87.

8. Ian A. McLaren, ed., *Natural Regulation of Animal Populations* (New York: Atherton Press, 1971), 116–35.

5. BEGINNING AN ACADEMIC CAREER

1. John A. Wiens, "Effects of Early Experience on Substrate Pattern Selection in *Rana aurora* Tadpoles," *Copeia* (1970): 543–48.

2. John A. Wiens, "Anuran Habitat Selection: Early Experience and Substrate Selection in *Rana cascadae* Tadpoles," *Animal Behaviour* 20 (1972): 218–20.

6. EXPANDING MY VIEW OF GRASSLAND BIRDS

1. Robert H. MacArthur and John W. MacArthur, "On Bird Species Diversity," *Ecology* 42 (1961): 594–98.

2. Robert H. MacArthur, letter to author, December 15, 1966.

7. EXTENDING THE PATHWAY: THE INTERNATIONAL BIOLOGICAL PROGRAM

1. John would continue working with me as a graduate student, and then a postdoctoral associate, for another decade as we attempted to make sense of the bird communities in grasslands and, later, in Great Basin shrubsteppe.
2. John A. Wiens, "Pattern and Process in Grassland Bird Communities," *Ecological Monographs* 43 (1973): 237–70.
3. Wiens, "Pattern and Process," 237–70.
4. G. Evelyn Hutchinson, "Homage to Santa Rosalia or Why are There so Many Kinds of Animals?," *The American Naturalist* 93 (1959): 145–59; see also John M. Eadie, Louis Broekhoven, and Patrick Colgan, "Size Ratios and Artifacts: Hutchinson's Rule Revisited," *The American Naturalist* 129 (1987): 1–17.
5. John A. Wiens and John T. Rotenberry, "Morphological Size Ratios and Competition in Ecological Communities," *The American Naturalist* 117 (1981): 592–99; John A. Wiens, "On Size Ratios and Sequences in Ecological Communities: Are There No Rules?," *Annales Zoologica Fennici* 19 (1982a): 297–308.
6. John A. Wiens and John T. Rotenberry, "Diet Niche Relationships Among North American Grassland and Shrubsteppe Birds," *Oecologia* 42 (1979): 253–92.

8. SCRAMBLING FOR AN EXPLANATION: CLIMATIC INSTABILITY AND ECOLOGICAL CRUNCHES

1. John A. Wiens, "On Competition and Variable Environments," *American Scientist* 65 (1977): 590–97; John A. Wiens, "Competition or Peaceful Coexistence?," *Natural History* 92, no. 3 (1983): 30–34.
2. John A. Wiens, "Climatic Instability and the 'Ecological Saturation' of Bird Communities in North American Grasslands," *The Condor* 76 (1974): 385–400.

9. DETOURING TO ANOTHER PATHWAY: MODELING BIRD BIOENERGETICS

1. For the details, see John A. Wiens, "Pattern and Process in Grassland Bird Communities," *Ecological Monographs* 43 (1973): 237–70; and John A. Wiens

and George S. Innis, "Estimation of Energy Flow in Bird Communities: A Population Bioenergetics Model," *Ecology* 55 (1974): 730–46.

2. Published by S. Charles Kendeigh, who I mentioned earlier.

3. John A. Wiens and Ronald A. Nussbaum, "Model Estimation of Energy Flow in Northwestern Coniferous Forest Bird Communities," *Ecology* 56 (1975): 547–61.

4. John A. Wiens and J. Michael Scott, "Model Estimation of Energy Flow in Oregon Coastal Seabird Populations," *The Condor* 77 (1975): 439–52; John A. Wiens, "Modeling the Energy Requirements of Seabird Populations," in *Seabird Energetics*, ed. G. Causey Whittow and Hermann Rahn (New York: Plenum Press, 1984), 255–84.

5. To put this in perspective, from 1966 to 1971 the commercial fishery operating in the area from Point Conception in Southern California north to the Oregon border took an average of 5,920 metric tons of anchovies per year, one-quarter of the estimated consumption by shearwaters.

6. John A. Wiens and Melvin I. Dyer, "Simulation Modeling of Red-Winged Blackbird Impact on Grain Crops," *Journal of Applied Ecology* 12 (1975): 63–82.

7. John A. Wiens and Melvin I. Dyer, "Assessing the Potential Impact of Granivorous Birds in Ecosystems," in *Granivorous Birds in Ecosystems*, ed. Jan Pinowski and S. Charles Kendeigh (Cambridge: Cambridge University Press, 1977), 205–66.

10. MOVING FROM GRASSLANDS TO THE ARID SHRUBSTEPPE

1. It turned out that sage sparrows had no special physiological abilities for living off of metabolic water; they had to make do with water obtained from their food or the environment. Which they did, just fine. See Ralph R. Moldenhauer and John A. Wiens, "The Water Economy of the Sage Sparrow, *Amphispiza belli nevadensis*," *The Condor* 72 (1970): 265–75.

2. As an aside, the IBP plots at ALE lay at the base of Rattlesnake Mountain—snakes again!

3. The origin of the name is uncertain. The last time there was a lake there was probably in the late Pleistocene. But now and then people did show up with fishing gear, asking "Where's the lake?"

4. John A. Wiens and John T. Rotenberry, "Habitat Associations and Community Structure of Birds in Shrubsteppe Environments," *Ecological Monographs* 51 (1981): 21–41.

5. John T. Rotenberry and John A. Wiens, "Temporal Variation in Habitat Structure and Shrubsteppe Bird Dynamics," *Oecologia* 47 (1980): 1–9.

6. A sampling method originally devised by John T. Emlen, "Population Densities of Birds Derived from Transect Counts," *The Auk* 88 (1971): 323–42.

7. William Shakespeare, *Henry IV, Part 1* (London: Andrew Wise, 1597): Act V, Scene 4.

8. John T. Rotenberry, "Habitat Relationships of Shrubsteppe Birds: Even 'Good' Models Cannot Predict the Future," in *Modeling Habitat Relationships of Terrestrial Vertebrates*, ed. J. Verner, M. L. Morrison, and C. J. Ralph (Madison: University of Wisconsin Press, 1986), 217–21.

9. John T. Rotenberry and John A. Wiens, "Habitat Relations of Shrubsteppe Birds: A 20-year Retrospective," *The Condor* 111 (2009): 401–13.

10. Rotenberry, "Habitat Relationships of Shrubsteppe Birds," 217–21.

11. Rotenberry, "Habitat Relationships of Shrubsteppe Birds," 217–21.

12. John A. Wiens, "Habitat Selection in Variable Environments: Shrub-Steppe Birds," in *Habitat Selection in Birds*, ed. Martin L. Cody (New York: Academic Press, 1985), 227–51.

13. Joseph H. Connell, "Diversity and the Coevolution of Competitors, or the Ghost of Competition Past," *Oikos* 35 (1980): 131–38.

14. John A. Wiens and John T. Rotenberry, "Patterns of Morphology and Ecology in Grassland and Shrubsteppe Bird Populations," *Ecological Monographs* 50 (1980): 287–308.

15. John T. Rotenberry, "Components of Avian Diversity Along a Multifactorial Gradient," *Ecology* 59 (1978): 693–99.

16. John A. Wiens, "On Understanding a Non-Equilibrium World: Myth and Reality in Community Patterns and Processes," in *Ecological Communities: Conceptual Issues and the Evidence*, ed. Donald R. Strong et al. (Princeton, NJ: Princeton University Press, 1984), 439–57.

11. CHALLENGING THE PARADIGM

1. John B. Dunning Jr., "Shrub-Steppe Bird Assemblages Revisited: Implications for Community Theory," *The American Naturalist* 128 (1986): 82–98; John A. Wiens and John T. Rotenberry, "Shrubsteppe Birds and the Generality of Community Models: A Response to Dunning," *The American Naturalist* 129 (1987): 920–27.

2. In ecology, a static equilibrium exists when the components of a system (e.g., abundances of species in a community) do not change; in a dynamic equilibrium, the components may change but their relationships to one another (or to resources such as food) remain constant.

3. Karl R. Popper, *The Logic of Scientific Discovery* (New York: Basic Books, 1959).

4. John A. Wiens, "On Competition and Variable Environments," *American Scientist* 65 (1977): 590–97, and "On Understanding a Non-Equilibrium World: Myth and Reality in Community Patterns and Processes," in *Ecological Communities: Conceptual Issues and the Evidence*, ed. Donald R. Strong Jr. et al. (Princeton, NJ: Princeton University Press, 1984), 439–57; see Nicholas J. Gotelli and Gary R. Graves, *Null Models in Ecology* (Washington, DC: Smithsonian Institution, 1996).
5. Andre A. Dhondt, *Interspecific Competition in Birds* (Oxford: Oxford University Press, 2012).
6. J. David Wiens, Robert G. Anthony, and Eric D. Forsman, "Competitive Interactions and Resource Partitioning between Northern Spotted Owls and Barred Owls in Western Oregon," *Wildlife Monographs* 185 (2014): 1–50; to add to the irony, the editor who handled the publication of David's research was Martin Cody.
7. John A. Wiens, *The Ecology of Bird Communities.* Vols. 1 and 2 (Cambridge: Cambridge University Press, 1989).

13. TESTING THE PARADIGM: ARE AUSTRALIAN BIRD COMMUNITIES DIFFERENT?

1. They are even dominated in many places by shrubs of the same genus, *Atriplex.*
2. John A. Wiens, "Ecological Similarity of Shrub-Desert Avifaunas of Australia and North America," *Ecology* 72 (1991): 479–95.
3. S. R. Morton, "Diversity of Desert-Dwelling Mammals: A Comparison of Australia and North America," *Journal of Mammalogy* 60 (1979): 253–64; Eric R. Pianka, *Ecology and Natural History of Desert Lizards* (Princeton, NJ: Princeton University Press, 1986).
4. John A. Wiens, "Ecomorphological Comparisons of the Shrub-Desert Avifaunas of Australia and North America," *Oikos* 60 (1991): 55–63.
5. Herbert G. Andrewartha and L. Charles Birch, *The Distribution and Abundance of Animals* (Chicago: University of Chicago Press, 1954).
6. For which he was awarded the prestigious Templeton Prize in 1990.

14. SHIFTING DIRECTIONS IN THE SHRUBSTEPPE

1. John A. Wiens, Beatrice Van Horne, and John T. Rotenberry, "Temporal and Spatial Variation in the Behavior of Shrubsteppe Birds," *Oecologia* 73 (1987): 60–70.

2. John A. Wiens, Beatrice Van Horne, and John T. Rotenberry, "Comparisons of the Behavior of Sage and Brewer's Sparrows in Shrubsteppe Habitat," *The Condor* 92 (1990): 264–66.

3. John T. Rotenberry and John A. Wiens, "Reproductive Biology of Shrubsteppe Passerine Birds: Geographical and Temporal Variation in Clutch Size, Brood Size, and Fledging Success," *The Condor* 91 (1989):1–14.

4. Rotenberry and Wiens, "Reproductive Biology," 1–14; John T. Rotenberry and John A. Wiens, "Weather and Reproductive Variation in Shrubsteppe Sparrows: A Hierarchical Analysis," *Ecology* 72 (1991): 1325–35.

5. Wiens, Van Horne, and Rotenberry, "Comparisons of the Behavior," 264–66.

6. John A. Wiens, "Song Pattern Variation in the Sage Sparrow (*Amphispiza belli*): Dialects or Epiphenomena?," *The Auk* 99 (1982): 208–29

15. FINDING THINGS OUT: FIELD EXPERIMENTS IN THE SHRUBSTEPPE

1. So named because of guano deposits that were found around the shores of a small lake in the valley, a vestigial remnant of the large lake that covered the valley to a depth of 30–40 m during the Pleistocene. Members of the U.S. Cavalry named the area in the mid-nineteenth century. The "lake" is now dry except in extraordinarily wet years.

2. John A. Wiens and John T. Rotenberry, "The Response of Breeding Passerine Birds to Rangeland Alteration in a North American Shrubsteppe Locality," *Journal of Applied Ecology* 22 (1985): 655–68.

3. The record was held by a male sage sparrow, Red-Over-White (his color leg-band combination), that we banded as an adult of unknown age who returned for the following seven summers to occupy the same territory.

4. John A. Wiens et al., "Arthropod Dynamics on Sagebrush (*Artemisia tridentata*): Effects of Plant Chemistry and Avian Predation," *Ecological Monographs* 61 (1991): 299–321.

16. BECOMING A LANDSCAPE ECOLOGIST

1. John A. Wiens, "Population Responses to Patchy Environments," *Annual Review of Ecology and Systematics* 7 (1976): 81–120.

2. Paul G. Risser, James R. Karr, and Richard T. T. Forman, *Landscape Ecology: Directions and Approaches*, Illinois Natural History Survey Special Publication, no. 2 (Champaign: Illinois Natural History Survey, 1984).

3. John A. Wiens, Clifford S. Crawford, and James R. Gosz, "Boundary Dynamics: A Conceptual Framework for Studying Landscape Ecosystems," *Oikos* 45 (1985): 421–27.
4. The paper we wrote summarizing our discussions [John A. Wiens et al., "Ecological Mechanisms and Landscape Ecology," *Oikos* 66 (1993): 369–80] was widely cited and received an award as the Best Paper in the Discipline of Landscape Ecology from the U.S. chapter of the International Association for Landscape Ecology in 1994. It helped to establish the credibility of landscape ecology as an ecological science.

17. DEALING WITH SCALE

1. John A. Wiens, "Spatial Scale and Temporal Variation in Studies of Shrub-steppe Birds," in *Community Ecology*, ed. Jared Diamond and Ted J. Case (New York: Harper & Row, 1986), 154–72; John A. Wiens, John T. Rotenberry, and Beatrice Van Horne, "Habitat Occupancy Patterns of North American Shrubsteppe Birds: The Effects of Spatial Scale," *Oikos* 48 (1987): 132–47.
2. John A. Wiens, "Scale Problems in Avian Censusing," in *Estimating the Numbers of Terrestrial Birds*, ed. C. John Ralph and J. Michael Scott, *Studies in Avian Biology* 6 (1981): 513–21.
3. Our discussions led to a joint publication conceptualizing hierarchical scales of patchiness, Natasha B. Kotliar and John A. Wiens, "Multiple Scales of Patchiness and Patch Structure: A Hierarchical Framework for the Study of Heterogeneity," *Oikos* 59 (1990): 253–60.
4. John A. Wiens, "Spatial Scaling in Ecology," *Functional Ecology* 3 (1989): 385–97.

18. FOLLOWING THE LANDSCAPE ECOLOGY PATHWAY

1. John A. Wiens and Bruce T. Milne, "Scaling of 'Landscapes' in Landscape Ecology, or Landscape Ecology from a Beetle's Perspective," *Landscape Ecology* 3 (1989): 87–96.
2. Benoit B. Mandelbrot's *The Fractal Geometry of Nature* (New York: Freeman Press, 1982) remains one of the best introductions to the world of fractals.
3. Wiens and Milne, "Scaling of 'Landscapes,' " 87–96.

4. T. O. Crist et al., "Animal Movements in Heterogeneous Landscapes: An Experiment with *Eleodes* Beetles in Shortgrass Prairie," *Functional Ecology* 6 (1992): 536–44.

5. Rodd Dyer et al., *Savanna Burning. Understanding and Using Fire in Northern Australia* (Darwin, NT, Australia: Tropical Savannas CRC, 2001); Ross A. Bradstock, Jann E. Williams, and A. Malcolm Gill, eds., *Flammable Australia. The Fire Regimes and Biodiversity of a Continent* (Cambridge: Cambridge University Press, 2002).

6. These dynamics are now changing rapidly as climate change accelerates.

7. Some, containing the remains of ancestors, were still visited regularly. George Chaloupka's *Journey in Time: The World's Longest Continuing Art Tradition* (Chatswood, NSW, Australia: Reed, 1993) provides an interesting overview of Aboriginal rock art in Arnhem Land.

8. John A. Wiens, "The Landscape Context of Dispersal," in *Dispersal: Individual, Population, and Community*, ed. Jean Clobert et al. (Oxford: Oxford University Press, 2001), 96–109.

9. John A. Wiens, "Central Concepts and Issues of Landscape Ecology," in *Applying Landscape Ecology in Biological Conservation*, ed. Kevin J. Gutzwiller (New York: Springer, 2002), 3–21.

10. I served as president of the International Association for Landscape Ecology (IALE) for four years and hosted the 5th World Congress of Landscape Ecology at Snowmass, Colorado, in 1999. Contributions to the World Congress resulted in a book: John Wiens and Michael Moss, eds., *Issues and Perspectives in Landscape Ecology* (Cambridge: Cambridge University Press, 2005).

19. TRAVELING ALONG PATHWAYS WITH STUDENTS

1. Some of whom had been part of the IBP Grasslands Biome program years before (see chapter 7).

2. Information about GDPE can be found at https://ecology.colostate.edu/.

3. John A. Wiens et al., "Fractal Patterns of Insect Movement in Microlandscape Mosaics," *Ecology* 76 (1995): 663–66.

4. Brandon T. Bestlemeyer and John A. Wiens, "The Effects of Land Use on the Structure of Ground-Foraging Ant Communities in the Argentine Chaco," *Ecological Applications* 6 (1996): 1225–40.

5. Brandon T. Bestlemeyer and John A. Wiens, "Ant Biodiversity in Semiarid Landscape Mosaics: The Consequences of Grazing vs. Natural Heterogeneity," *Ecological Applications* 11 (2001): 1123–40.

6. Later to be consolidated into the Midcontinent Ecological Science Center, and now the Fort Collins Science Center of USGS, still in Fort Collins, CO.

7. In dynamic programming, a computer program that addresses a general problem is broken down into a series of subprograms, which are individually optimized and then combined to give an overall solution.

8. Adrian H. Farmer and John A. Wiens, "Optimal Migration Schedules Depend on the Landscape and the Physical Environment: A Dynamic Modeling View," *Journal of Avian Biology* 29 (1998): 405–15, and "Models and Reality: Time-Energy Tradeoffs in Pectoral Sandpiper (*Calidris melanotos*) Migration," *Ecology* 80 (1999): 2566–80.

20. PULLED ONTO A SEABIRD PATHWAY

1. As, in fact, Martin Cody had also done in a less empirical, more conceptual study; Martin L. Cody, "Coexistence, Coevolution, and Convergent Evolution in Seabird Communities," *Ecology* 54 (1973): 31–44.

2. So named by a fur trader when a party he sent ashore for water and timber in the late eighteenth century was massacred by Indigenous people.

3. It's never been clear to me what was philosophical about our project, but I was grateful for their support.

4. It is now part of the Quillayute Needles National Wildlife Refuge.

5. Wayne Hoffman, John A. Wiens, and J. Michael Scott, "Hybridization Between Gulls (*Larus glaucescens* and *L. occidentalis*) in the Pacific Northwest," *The Auk* 95 (1978): 441–58.

6. Wayne Hoffman, Dennis Heinemann, and John A. Wiens, "The Ecology of Seabird Feeding Flocks in Alaska," *The Auk* 98 (1981): 437–56.

7. R. G. Ford et al., "Modeling the Sensitivity of Colonially Breeding Marine Birds to Oil Perturbations: Guillemot and Kittiwake Populations on the Pribilof Islands, Bering Sea," *Journal of Applied Ecology* 19 (1982): 1–31.

21. DEALING WITH ADVOCACY: THE *EXXON VALDEZ* OIL SPILL

1. According to detailed modeling conducted by Glenn Ford. John F. Piatt and R. Glenn Ford, "How Many Seabirds were Killed by the *Exxon Valdez* Oil Spill," in *Proceedings of the* Exxon Valdez *Oil Spill Symposium*, ed. Stanley D.

Rice et al. (Bethesda, MD: American Fisheries Society Symposium 18, 1996), 712–19.

2. Later to become ExxonMobil. I refer to it as "Exxon" throughout this chapter, as that is what it was called when I first became involved.

3. See John A. Wiens and Keith R. Parker, "Analyzing the Effects of Accidental Environmental Impacts: Approaches and Assumptions," *Ecological Applications* 5 (1995): 1069–83; and Keith R. Parker and John A. Wiens, "Assessing Environmental Accidents: Environmental Variation, Ecological Assumptions, and Strategies," *Ecological Applications* 15 (2005): 2037–51.

4. John A. Wiens et al., "Effects of the *Exxon Valdez* Oil Spill on Marine Bird Communities in Prince William Sound, Alaska," *Ecological Applications* 6 (1996): 828–41.

5. Robert H. Day et al., "Effects of the *Exxon Valdez* Oil Spill on Habitat Use by Birds in Prince William Sound, Alaska," *Ecological Applications* 7 (1997): 593–613.

6. John A. Wiens et al., "Changing Habitat and Habitat Use by Birds after the *Exxon Valdez* Oil Spill, 1989–2001," *Ecological Applications* 14 (2004): 1806–25.

7. The last remaining lawsuit was settled in 2015.

8. John A. Wiens, ed., *Oil in the Environment: Lessons and Legacies of the* Exxon Valdez *Oil Spill* (Cambridge: Cambridge University Press, 2013).

22. SHIFTING PATHWAYS TO CONSERVATION

1. Holmes Rolston III, *Conserving Natural Value* (New York: Columbia University Press, 1994), and *A New Environmental Ethics: The Next Millennium for Life on Earth* (New York: Routledge, 2012); and Ronald L. Sandler, *The Ethics of Species: An Introduction* (Cambridge: Cambridge University Press, 2012) all provide cogent discussions of the intrinsic and instrumental values of nature.

2. John A. Wiens, "Habitat Fragmentation and Wildlife Populations: The Importance of Autecology, Time, and Landscape Structure," *Proceedings: 19th International Union of Game Biologists Congress, Trondheim* 1989 (1990): 381–91.

3. Now renamed the Department of Fish, Wildlife, and Conservation Biology.

4. The "bucks and acres" approach to conservation.

5. Virginia Dale, Frank Davis, John Ewel, Malcolm Hunter, John Ogden, Mary Power, and Margaret Shannon.

23. BOLSTERING CONSERVATION SCIENCE IN THE NATURE CONSERVANCY

1. Sanjayan is from Sri Lanka, where people often go by a single name.
2. Climate change could also enhance the conservation value of some places. I once suggested that we should target some damaged and degraded places for conservation with the expectation that they would provide future protection for species displaced by climate change, which would increase their conservation value. My suggestion generated little enthusiasm—an organization that promoted itself as protecting pretty places couldn't be seen by donors as investing in trashy places.
3. Within a few years, however, climate change had been incorporated into many TNC programs and it now figures prominently in their actions and publicity.
4. Life in the Washington, DC, metropolitan area was much more closely aligned with nine-to-five working hours than in the more loosely organized and informal setting of academia, perhaps because most of us faced long commutes home at the end of the day.
5. Greg had previously been a postdoc working with me on assessing the consequences of the *Exxon Valdez* oil spill.
6. The results of the workshop were published in an edited volume: John A. Wiens et al., eds., *Historical Environmental Variation in Conservation and Natural Resource Management* (Oxford: Wiley-Blackwell, 2012).
7. The bloodiest battle in American military history.

24. DOING CONSERVATION SCIENCE

1. Nadav Nur et al., "Where the Wild Things Are: Predicting Hotspots of Seabird Aggregations in the California Current System," *Ecological Applications* 21 (2011): 2241–57.
2. Diana Stralberg et al., "Re-Shuffling of Species with Climate Disruption: A No-Analog Future for California Birds?," *PLoS One* 4, no. 9 (2009): 6825, edoc 10.1371/, https://journals.plos.org/plosone/article?id=10.1371/journal.pone.0006825.
3. Dennis Jongsomjit et al., "Between a Rock and a Hard Place: The Impacts of Climate Change and Housing Development on Breeding Birds in California," *Landscape Ecology* DOI 10.1007/s10980-012-9825-1 (2012).
4. For maps illustrating this example, see J. Michael Scott et al., *Shepherding Nature: The Challenge of Conservation Reliance* (Cambridge: Cambridge University Press, 2020), 148.

5. This is perhaps why Bubba Farmer and I were unable to obtain NSF funding for continuing his shorebird studies, as I describe in chapter 19.
6. John A. Wiens et al., "Niches, Models, and Climate Change: Assessing the Assumptions and Uncertainties," *PNAS* 106 (Suppl. 2, 2009): 19729–36.

25. WRITING INTO RETIREMENT

1. Ornithological societies have often named their journals after charismatic birds—not just *The Auk* (American Ornithologists' Union) but *The Condor* (Cooper Ornithological Society), *The Ibis* (British Ornithologists' Union), *The Emu* (BirdLife Australia), *The Ostrich* (BirdLife South Africa), and others.
2. At least anything potentially interesting to ecologists, who are interested in pretty much everything.
3. Richard and I also wrote about the differences in a contribution to a book about the history of landscape ecology in the United States: John A. Wiens and Richard J. Hobbs, "A Tale of Two Continents: The Growth and Maturation of Landscape Ecology in North America and Australia," in *History of Landscape Ecology in the United States*, ed. Gary W. Barrett, Terry L. Barrett, and Jianguo Wu (New York: Springer. 2015), 143–61.
4. John A. Wiens, *Ecological Challenges and Conservation Conundrums. Essays and Reflections for a Changing World* (Chichester, UK: John Wiley & Sons, 2016).
5. J. Michael Scott et al., "Recovery of Imperiled Species Under the Endangered Species Act: The Need for a New Approach," *Frontiers in Ecology and the Environment* 3 (2005): 383–89; J. Michael Scott et al., "Conservation-Reliant Species and the Future of Conservation," *Conservation Letters* 3 (2010): 91–97; Dale D. Goble et al., "Conservation-Reliant Species," *BioScience* 62 (2012): 869–73; John A. Wiens and Thomas Gardali, "Conservation Reliance Among California's At-Risk Birds," *The Condor* 115, no. 3 (2013): 1–15.
6. J. Michael Scott et al., *Shepherding Nature. The Challenge of Conservation Reliance* (Cambridge: Cambridge University Press, 2020).
7. This has led to the metric of the Least Publishable Unit (LPU), an expression of the tendency to subdivide the results of a single research project into as many separate publications as possible. This may impress those who count rather than read scientific papers (e.g., a university dean I once knew), but it fragments the results of a study and erodes any cohesion and complementarity they might otherwise have.

CONCLUSIONS: WHAT ARE THE LESSONS FOR TODAY'S ASPIRING ECOLOGISTS?

1. Rudyard Kipling, *Just So Stories for Little Children* (London: Macmillan, 1902).
2. When I taught an introductory ecology class, I sometimes took the students out to the woods and told them to go off, find a nice place to sit, observe what they saw for half an hour, and come back with a question or two to share with the group. It was my favorite lab exercise, an attempt to teach them the art of asking questions. At least some of them didn't find it boring, and several developed research projects as a result.
3. I long argued, to no avail, that the training of an ecologist—or any scientist—should include a course in ethics.

BIBLIOGRAPHY

Andrewartha, Herbert G., and L. Charles Birch. *The Distribution and Abundance of Animals*. Chicago: University of Chicago Press, 1954.

Bestlemeyer, Brandon T., and John A. Wiens. "Ant Biodiversity in Semiarid Landscape Mosaics: The Consequences of Grazing vs. Natural Heterogeneity." *Ecological Applications* 11 (2001): 1123–40.

——. "The Effects of Land Use on the Structure of Ground-Foraging Ant Communities in the Argentine Chaco." *Ecological Applications* 6 (1996): 1225–40.

Bradstock, Ross A., Jann E. Williams, and A. Malcolm Gill, eds. *Flammable Australia: The Fire Regimes and Biodiversity of a Continent*. Cambridge: Cambridge University Press, 2002.

Chaloupka, George. *Journey in Time: The World's Longest Continuing Art Tradition*. Chatswood, NSW, Australia: Reed, 1993.

Cody, Martin L. "Coexistence, Coevolution, and Convergent Evolution in Seabird Communities." *Ecology* 54 (1973): 31–44.

Connell, Joseph H. "Diversity and the Coevolution of Competitors, or the Ghost of Competition Past." *Oikos* 35 (1980): 131–38.

Crist, Thomas O., David S. Guertin, John A. Wiens, and Bruce T. Milne. "Animal Movements in Heterogeneous Landscapes: An Experiment with *Eleodes* Beetles in Shortgrass Prairie." *Functional Ecology* 6 (1992): 536–44.

Cuddington, Kim, and Beatrix Beisner, eds. *Ecological Paradigms Lost: Routes of Theory Change*. Burlington, MA: Elsevier Academic Press, 2005.

Day, Robert H., Stephen M. Murphy, John A. Wiens, Gregory D. Hayward, E. James Harner, and Louise N. Smith. "Effects of the *Exxon Valdez* Oil Spill on Habitat

Use by Birds in Prince William Sound, Alaska." *Ecological Applications* 7 (1997): 593–613.

Dhondt, Andre A. *Interspecific Competition in Birds.* Oxford: Oxford University Press, 2012.

Dunning, John B., Jr. "Shrub-steppe Bird Assemblages Revisited: Implications for Community Theory." *The American Naturalist* 128 (1986): 82–98.

Dyer, Rodd, Peter Jacklyn, Ian Partridge, Jeremy Russell-Smith, and Dick Williams. *Savanna Burning: Understanding and Using Fire in Northern Australia.* Darwin, NT, Australia: Tropical Savannas CRC, 2001.

Eadie, John M., Louis Broekhoven, and Patrick Colgan. "Size Ratios and Artifacts: Hutchinson's Rule Revisited." *The American Naturalist* 129 (1987): 1–17.

Emlen, John T. "Population Densities of Birds Derived from Transect Counts." *The Auk* 88 (1971): 323–42.

Errington, Paul L. *Of Men and Marshes.* Ames: Iowa State University Press, 1957.

Farmer, Adrian H., and John A. Wiens. "Models and Reality: Time-Energy Trade-offs in Pectoral Sandpiper (*Calidris melanotos*) Migration." *Ecology* 80 (1999): 2566–80.

——. "Optimal Migration Schedules Depend on the Landscape and the Physical Environment: A Dynamic Modeling View." *Journal of Avian Biology* 29 (1998): 405–15.

Feynman, Richard P. *The Pleasure of Finding Things Out.* Cambridge, MA: Perseus Books, 1999.

Ford, R. Glenn, John A. Wiens, Dennis Heinemann, and George L. Hunt. "Modeling the Sensitivity of Colonially Breeding Marine Birds to Oil Perturbations: Guillemot and Kittiwake Populations on the Pribilof Islands, Bering Sea." *Journal of Applied Ecology* 19 (1982): 1–31.

Goble, Dale D., John A. Wiens, J. Michael Scott, Timothy D. Male, and John A. Hall. "Conservation-Reliant Species." *BioScience* 62 (2012): 869–73.

Gotelli, Nicholas J., and Gary R. Graves. *Null Models in Ecology.* Washington, DC: Smithsonian Institution, 1996.

Hardin, Garrett. "Human Ecology: The Subversive, Conservative Science." *American Zoologist* 25 (1985): 469–76.

Hoffman, Wayne, John A. Wiens, and J. Michael Scott. "Hybridization Between Gulls (*Larus glaucescens* and *L. occidentalis*) in the Pacific Northwest." *The Auk* 95 (1978): 441–58.

Hoffman, Wayne, Dennis Heinemann, and John A. Wiens. "The Ecology of Seabird Feeding Flocks in Alaska." *The Auk* 98 (1981): 437–56.

Hunter, Malcolm L., David B. Lindenmayer, and Aram J. K. Calhoun. *Saving the Earth as a Career: Advice on Becoming a Conservation Professional, Second Edition.* Chichester, UK: John Wiley & Sons, 2016.

Hutchinson, G. Evelyn. "Homage to Santa Rosalia or Why are There so Many Kinds of Animals?" *The American Naturalist* 93 (1959): 145–59.

Jackson, Jerome A. *George Miksch Sutton: Artist, Scientist, and Teacher*. Norman: University of Oklahoma Press, 2007.

Jongsomjit, Dennis, Diana Stralberg, Thomas Gardali, Leonardo Salas, and John A. Wiens. "Between a Rock and a Hard Place: The Impacts of Climate Change and Housing Development on Breeding Birds in California." *Landscape Ecology* DOI 10.1007/s10980-012-9825-1 (2012).

Juavinett, Ashley. *So You Want to Be a Neuroscientist?* New York: Columbia University Press, 2020.

Kipling, Rudyard. *Just So Stories for Little Children*. London: Macmillan, 1902.

Kotliar, Natasha B., and John A. Wiens. "Multiple Scales of Patchiness and Patch Structure: A Hierarchical Framework for the Study of Heterogeneity." *Oikos* 59 (1990): 253–60.

Kuhn, Thomas S. *The Structure of Scientific Revolutions, Second Edition*. Chicago: University of Chicago Press, 1970.

Lakatos, Imre, and Alan Musgrave, eds. *Criticism and the Growth of Knowledge*. London: Cambridge University Press, 1970.

Leopold, Aldo. *Round River*, edited by Luna B. Leopold, 173. New York: Oxford University Press, 1953.

Louv, Richard. *Last Child in the Woods: Saving our Children from Nature-Deficit Disorder*. London: Atlantic Books, 2005.

MacArthur, Robert H. "Population Ecology of Some Warblers of Northeastern Coniferous Forests." *Ecology* 39 (1958): 599–619.

MacArthur, Robert H., and John W. MacArthur. "On Bird Species Diversity." *Ecology* 42 (1961): 594–98.

Malthus, Thomas Robert. *An Essay on the Principle of Population*. London: J. Johnson, 1798.

Mandelbrot, Benoit B. *The Fractal Geometry of Nature*. New York: Freeman Press, 1982.

McLaren, Ian A., ed. *Natural Regulation of Animal Populations*. New York: Atherton Press, 1971.

Moldenhauer, Ralph R., and John A. Wiens. "The Water Economy of the Sage Sparrow, *Amphispiza belli nevadensis*." *The Condor* 72 (1970): 265–75.

Morton, S. R. "Diversity of Desert-Dwelling Mammals: A Comparison of Australia and North America." *Journal of Mammalogy* 60 (1979): 253–64.

Nice, Margaret Morse. "Studies in the Life History of the Song Sparrow, Volume 1, *A Population Study of the Song Sparrow*." *Transactions of the Linnean Society of New York* 4 (1937): 1–247.

Nickles, Thomas, ed. *Thomas Kuhn*. Cambridge: Cambridge University Press, 2003.

Nur, Nadav, Jaime Jahncke, Mark P. Herzog, Julie Howar, K. David Hyrenbach, Jeanette E. Zamon, David G. Ainley, John A. Wiens, Ken Morgan, Lisa T. Balance, and Diana Stralberg. "Where the Wild Things Are: Predicting Hotspots of Seabird Aggregations in the California Current System." *Ecological Applications* 21 (2011): 2241–57.

Odum, Eugene P. *Fundamentals of Ecology, Second Edition.* Philadelphia: W. B. Saunders, 1959.

Ogilvie, Marilyn Bailey. *For the Birds: American Ornithologist Margaret Morse Nice.* Norman: University of Oklahoma Press, 2018.

Parker, Keith R., and John A. Wiens. "Assessing Environmental Accidents: Environmental Variation, Ecological Assumptions, and Strategies." *Ecological Applications* 15 (2005): 2037–51.

Pianka, Eric R. *Ecology and Natural History of Desert Lizards.* Princeton, NJ: Princeton University Press, 1986.

Piatt, John F., and R. Glenn Ford. "How Many Seabirds Were Killed by the *Exxon Valdez* Oil Spill." In *Proceedings of the* Exxon Valdez *Oil Spill Symposium*, edited by Stanley D. Rice, Robert B. Spies, Douglas A. Wolfe, and Bruce A. Wright, 712–19. Bethesda, MD: American Fisheries Society Symposium 18, 1996.

Plotnick, Roy. *Explorers of Deep Time. Paleontologists and the History of Life.* New York: Columbia University Press, 2022.

Popper, Karl R. *The Logic of Scientific Discovery.* New York: Basic Books, 1959.

Risser, Paul G., James R. Karr, and Richard T. T. Forman. *Landscape Ecology: Directions and Approaches.* Illinois Natural History Survey Special Publication, no. 2. Champaign: Illinois Natural History Survey, 1984.

Rolston, Holmes III. *A New Environmental Ethics: The Next Millennium for Life on Earth.* New York: Routledge, 2012.

——. *Conserving Natural Value.* New York: Columbia University Press, 1994.

Rotenberry, John T. "Components of Avian Diversity Along a Multifactorial Gradient." *Ecology* 59 (1978): 693–99.

——. "Habitat Relationships of Shrubsteppe Birds: Even 'Good' Models Cannot Predict the Future." In *Modeling Habitat Relationships of Terrestrial Vertebrates*, edited by Jared Verner, Michael L. Morrison, and C. John Ralph, 217–21. Madison: University of Wisconsin Press, 1986.

Rotenberry, John T., and John A. Wiens. "Habitat Relations of Shrubsteppe Birds: A 20-Year Retrospective." *The Condor* 111 (2009): 401–13.

——. "Reproductive Biology of Shrubsteppe Passerine Birds: Geographical and Temporal Variation in Clutch Size, Brood Size, and Fledging Success." *The Condor* 91 (1989): 1–14.

——. "Temporal Variation in Habitat Structure and Shrubsteppe Bird Dynamics." *Oecologia* 47 (1980): 1–9.

——. "Weather and Reproductive Variation in Shrubsteppe Sparrows: A Hierarchical Analysis." *Ecology* 72 (1991): 1325–35.

Sandler, Ronald L. *The Ethics of Species. An Introduction.* Cambridge: Cambridge University Press, 2012.

Schoener, Thomas W. ["Reply to Wiens 1983"]. *American Scientist* 71 (1983): 235.

Scott, J. Michael, Dale D. Goble, John A. Wiens, David S. Wilcove, Michael Bean, and Timothy Male. "Recovery of Imperiled Species Under the Endangered Species Act: The Need for a New Approach." *Frontiers in Ecology and the Environment* 3 (2005): 383–89.

Scott, J. Michael, Dale D. Goble, Aaron M. Haines, John A. Wiens, and Maile C. Neel. "Conservation-Reliant Species and the Future of Conservation." *Conservation Letters* 3 (2010): 91–97.

Scott, J. Michael, John A. Wiens, Beatrice Van Horne, and Dale D. Goble. *Shepherding Nature: The Challenge of Conservation Reliance.* Cambridge: Cambridge University Press, 2020.

Sears, Paul B. "Ecology—a Subversive Subject." *BioScience* 14, no. 7 (1964): 11–13.

Shakespeare, William. *Henry IV, Part 1.* London: Andrew Wise, 1597.

Stralberg, Diana, Dennis Jongsomjit, Christine A. Howell, Mark A. Snyder, John D. Alexander, John A. Wiens, and Terry L. Root. "Re-Shuffling of Species with Climate Disruption: A No-Analog Future for California Birds?" *PLoS One* 4, no. 9 (2009): 6825. edoc 10.1371. https://journals.plos.org/plosone/article?id=10.1371/journal.pone.0006825.

Sutton, George Miksch. *Oklahoma Birds: Their Ecology and Distribution, with Comments on the Avifauna of the Southern Great Plains.* Norman: University of Oklahoma Press, 1967.

Wiens, John A. "An Approach to the Study of Ecological Relationships Among Grassland Birds." *Ornithological Monographs* 8 (1969): 89.

——. "Anuran Habitat Selection: Early Experience and Substrate Selection in *Rana cascadae* Tadpoles." *Animal Behaviour* 20 (1972): 218–20.

——. "Aspects of Cowbird Parasitism in Southern Oklahoma." *Wilson Bulletin* 75 (1963): 130–39.

——. "Behavioral Interactions of Red-Winged Blackbirds and Common Grackles on a Common Breeding Ground." *The Auk* 82 (1965): 356–74.

——. "Climatic Instability and the 'Ecological Saturation' of Bird Communities in North American Grasslands." *The Condor* 76 (1974): 385–400.

——. "Competition or Peaceful Coexistence?" *Natural History* 92, no. 3 (1983): 30–34.

——. *Ecological Challenges and Conservation Conundrums. Essays and Reflections for a Changing World.* Chichester, UK: John Wiley & Sons, 2016.

——. "Ecological Similarity of Shrub-Desert Avifaunas of Australia and North America." *Ecology* 72 (1991): 479–95.

——. *The Ecology of Bird Communities, Vols. 1 & 2.* Cambridge: Cambridge University Press, 1989.

——. "Ecomorphological Comparisons of the Shrub-Desert Avifaunas of Australia and North America." *Oikos* 60 (1991): 55–63.

——. "Effects of Early Experience on Substrate Pattern Selection in *Rana aurora* Tadpoles." *Copeia* (1970): 543–48.

——. "Habitat Fragmentation and Wildlife Populations: The Importance of Autecology, Time, and Landscape Structure." *Proceedings: 19th International Union of Game Biologists Congress, Trondheim 1989* (1990): 381–91.

——. "Habitat Selection in Variable Environments: Shrub-Steppe Birds." In *Habitat Selection in Birds*, edited by Martin L. Cody, 227–51. New York: Academic Press, 1985.

——. "Interspecific Competition." *American Scientist* 71 (1983): 234–35.

——. "Interterritorial Habitat Variation in Grasshopper and Savannah Sparrows." *Ecology* 54 (1973): 877–84.

——. "The Landscape Context of Dispersal." In *Dispersal: Individual, Population, and Community*, edited by Jean Clobert, Etienne Danchin, André A. Dhondt, and James D. Nichols, 96–109. Oxford: Oxford University Press, 2001.

——. "Modeling the Energy Requirements of Seabird Populations." In *Seabird Energetics*, edited by G. Causey Whittow and Hermann Rahn, 255–84. New York: Plenum Press, 1984.

——. "On Competition and Variable Environments." *American Scientist* 65 (1977): 590–97.

——. "On Group Selection and Wynne-Edwards' Hypothesis." *American Scientist* 54 (1966): 273–87.

——. "On Size Ratios and Sequences in Ecological Communities: Are There no Rules?" *Annales Zoologica Fennici* 19 (1982): 297–308.

——. "On Understanding a Non-Equilibrium World: Myth and Reality in Community Patterns and Processes." In *Ecological Communities: Conceptual Issues and the Evidence*, edited by Donald R. Strong, Jr., Daniel Simberloff, Lawrence G. Abele, and Anne B. Thistle, 439–57. Princeton, NJ: Princeton University Press, 1984.

——. "Pattern and Process in Grassland Bird Communities." *Ecological Monographs* 43 (1973): 237–70.

——. "Population Responses to Patchy Environments." *Annual Review of Ecology and Systematics* 7 (1976): 81–120.

——. "Scale Problems in Avian Censusing." In *Estimating the Numbers of Terrestrial Birds*, edited by C. John Ralph and J. Michael Scott, 513–21. *Studies in Avian Biology* 6 (1981).

——. "Song Pattern Variation in the Sage Sparrow (*Amphispiza belli*): Dialects or Epiphenomena?" *The Auk* 99 (1982): 208–29.

——. "Spatial Scale and Temporal Variation in Studies of Shrubsteppe Birds." In *Community Ecology*, edited by Jared Diamond and Ted J. Case, 154–72. New York: Harper & Row, 1986.

——. "Spatial Scaling in Ecology." *Functional Ecology* 3 (1989): 385–97.

Wiens, John A., ed. *Oil in the Environment: Lessons and Legacies of the* Exxon Valdez *Oil Spill.* Cambridge: Cambridge University Press, 2013.

Wiens, John A., and Melvin I. Dyer. "Assessing the Potential Impact of Granivorous Birds in Ecosystems." In *Granivorous Birds in Ecosystems*, edited by Jan Pinowski and S. Charles Kendeigh, 205–66. Cambridge: Cambridge University Press, 1977.

——. "Simulation Modeling of Red-Winged Blackbird Impact on Grain Crops." *Journal of Applied Ecology* 12 (1975): 63–82.

Wiens, John A., and Thomas Gardali. "Conservation Reliance Among California's At-Risk Birds." *The Condor* 115, no. 3 (2013): 1–15.

Wiens, John A., and Richard J. Hobbs. "A Tale of Two Continents: The Growth and Maturation of Landscape Ecology in North America and Australia." In *History of Landscape Ecology in the United States*, edited by Gary W. Barrett, Terry L. Barrett, and Jianguo Wu, 143–61. New York: Springer, 2015.

Wiens, John A., and George S. Innis. "Estimation of Energy Flow in Bird Communities: A Population Bioenergetics Model." *Ecology* 55 (1974): 730–46.

Wiens, John A., and Charles Krebs. "Last Ecologist in the Woods?" *Bulletin of the British Ecological Society* 48, no. 2 (2017): 47–49.

Wiens, John A., and Bruce T. Milne. "Scaling of 'Landscapes' in Landscape Ecology, or Landscape Ecology from a Beetle's Perspective." *Landscape Ecology* 3 (1989): 87–96.

Wiens, John A., and Ronald A. Nussbaum. "Model Estimation of Energy Flow in Northwestern Coniferous Forest Bird Communities." *Ecology* 56 (1975): 547–61.

Wiens, John A., and Keith R. Parker. "Analyzing the Effects of Accidental Environmental Impacts: Approaches and Assumptions." *Ecological Applications* 5 (1995): 1069–83.

Wiens, John A., and John T. Rotenberry. "Diet Niche Relationships Among North American Grassland and Shrubsteppe Birds." *Oecologia* 42 (1979): 253–92.

——. "Habitat Associations and Community Structure of Birds in Shrubsteppe Environments." *Ecological Monographs* 51 (1981): 21–41.

——. "Morphological Size Ratios and Competition in Ecological Communities." *The American Naturalist* 117 (1981): 592–99.

——. "Patterns of Morphology and Ecology in Grassland and Shrubsteppe Bird Populations." *Ecological Monographs* 50 (1980): 287–308.

——. "The Response of Breeding Passerine Birds to Rangeland Alteration in a North American Shrubsteppe Locality." *Journal of Applied Ecology* 22 (1985): 655–68.

——. "Shrubsteppe Birds and the Generality of Community Models: A Response to Dunning." *The American Naturalist* 129 (1987): 920–27.

Wiens, John A., and J. Michael Scott. "Model Estimation of Energy Flow in Oregon Coastal Seabird Populations." *The Condor* 77 (1975): 439–52.

Wiens, John A., Clifford S. Crawford, and James R. Gosz. "Boundary Dynamics: A Conceptual Framework for Studying Landscape Ecosystems." *Oikos* 45 (1985): 421–27.

Wiens, John A., Beatrice Van Horne, and John T. Rotenberry. "Comparisons of the Behavior of Sage and Brewer's Sparrows in Shrubsteppe Habitat." *The Condor* 92 (1990): 264–66.

——. "Temporal and Spatial Variation in the Behavior of Shrubsteppe Birds." *Oecologia* 73 (1987): 60–70.

Wiens, John A., Rex G. Cates, John T. Rotenberry, Neil Cobb, Beatrice Van Horne, and Richard A. Redak. "Arthropod Dynamics on Sagebrush (*Artemisia tridentata*): Effects of Plant Chemistry and Avian Predation." *Ecological Monographs* 61 (1991): 299–321.

Wiens, John A., Nils Chr. Stenseth, Beatrice Van Horne, and Rolf A. Ims. "Ecological Mechanisms and Landscape Ecology." *Oikos* 66 (1993): 369–80.

Wiens, John A., Thomas O. Crist, Kimberly A. With, and Bruce T. Milne. "Fractal Patterns of Insect Movement in Microlandscape Mosaics." *Ecology* 76 (1995): 663–66.

Wiens, John A., Thomas O. Crist, Robert H. Day, Stephen M. Murphy, and Gregory D. Hayward. "Effects of the *Exxon Valdez* Oil Spill on Marine Bird Communities in Prince William Sound, Alaska." *Ecological Applications* 6 (1996): 828–41.

Wiens, John A., Robert H. Day, Stephen M. Murphy, and Keith R. Parker. "Changing Habitat and Habitat Use by Birds after the *Exxon Valdez* Oil Spill, 1989–2001." *Ecological Applications* 14 (2004): 1806–25.

Wiens, John A., Diana Stralberg, Dennis Jongsomjit, Christine A. Howell, and Mark A. Snyder. "Niches, Models, and Climate Change: Assessing the Assumptions and Uncertainties." *PNAS* 106 (Suppl. 2, 2009): 19729–36.

Wiens, John A., Gregory D. Hayward, Hugh D. Safford, and Catherine M. Giffen, eds. *Historical Environmental Variation in Conservation and Natural Resource Management*. Oxford: Wiley-Blackwell, 2012.

Wiens, J. David, Robert G. Anthony, and E. D. Forsman. "Competitive Interactions and Resource Partitioning Between Northern Spotted Owls and Barred Owls in Western Oregon." *Wildlife Monographs* 185 (2014): 1–50.

Wilson, E. O. *Naturalist*. Washington, DC: Island Press, 1994.

Wynne-Edwards, V. C. *Animal Dispersion in Relation to Social Behaviour*. New York: Hafner, 1962.

INDEX

GPSR Authorized Representative: Easy Access System Europe, Mustamäe tee
50, 10621 Tallinn, Estonia, gpsr.requests@easproject.com

www.ingramcontent.com/pod-product-compliance
Lightning Source LLC
Chambersburg PA
CBHW030457210326
41597CB00013B/702